T0295468

MARINE IMPACTS OF SEAWATER DESALINATION

MARINE IMPACTS OF SEAWATER DESALINATION

Science, Management, and Policy

NURIT KRESS
Senior Scientist, Israel Oceanographic and Limnological Research, National Institute of Oceanography, Haifa, Israel

Elsevier
Radarweg 29, PO Box 211, 1000 AE Amsterdam, Netherlands
The Boulevard, Langford Lane, Kidlington, Oxford OX5 1GB, United Kingdom
50 Hampshire Street, 5th Floor, Cambridge, MA 02139, United States

Library of Congress Cataloging-in-Publication Data
A catalog record for this book is available from the Library of Congress

British Library Cataloguing-in-Publication Data
A catalogue record for this book is available from the British Library

ISBN: 978-0-12-811953-2

For information on all Elsevier publications visit our website at https://www.elsevier.com/books-and-journals

Publisher: Candice Janco
Acquisition Editor: Louisa Munro
Editorial Project Manager: Michelle Fisher
Production Project Manager: Nilesh Kumar Shah
Cover designer: Mark Rogers

Typeset by SPi Global, India

To my husband and partner, Moshe
and to my daughters,
Hadas, Naama, and Noa.

CONTENTS

PREFACE

The idea to write this book was suggested to me by Elsevier, following the Ocean Sciences Meeting (OSM) that took place in New Orleans in 2016. At the OSM I chaired, with Prof. Ilana Berman-Frank, a dedicated section on "The impacts of seawater desalination on the marine and coastal environment" (ID# 9245). The motivation for this section was to introduce the topic to the ocean research community, point out the lack of data on the actual impacts, and stimulate more involvement in research.

Seawater desalination is increasing globally in conjunction with the increased demand for fresh water supply. The potential effects of seawater desalination on the marine environment are detailed and explained in many publications. Common, but inaccurate, "wisdom" states that if properly engineered, seawater desalination will have no impact. The problem is that this view is currently not backed by data. Actual observations on environmental and ecological impacts of desalination plants on the marine environment are scarce even now in 2018.

Notwithstanding the above limitations, this book presents the current state of knowledge on this complex and environmentally critical topic. It contains a compilation and review of existing observational data on the actual effects of seawater desalination on the marine environment. Those data and observations were published in peer-reviewed literature, environmental impact assessment and monitoring reports, and conference proceedings or conveyed by personal communications. Moreover, the findings are critically reviewed, and most important, knowledge gaps are identified and emphasized.

This book is multidisciplinary in character and targets a diverse community: engineers and developers in the desalination industry can use this book to guide technological development toward mitigation and prevention of environmental impact; researchers can tune their laboratory and in situ studies to provide new data and reduce knowledge gaps; regulators and decision makers can use the data and the tools provided to set up guidelines for the protection of the marine environment and to help with permitting for plant operations; and students and teachers in environmental studies programs will profit from the data collation and the multidisciplinary approach to environmental issues.

This book comprises seven chapters as outlined in this preface. Seawater desalination is emphasized in all of them. Chapters 2, 3, and 7, covering necessary ancillary topics with extensive research, are brief and include references for additional information.

Chapter 1 presents an overview of topics that, although associated with seawater desalination, are not covered in this book. They include identifying the need for seawater desalination, through a short history of the process, global water desalination market and policy, the water-energy-food-ecosystem nexus, and the quality of desalinated water.

Chapter 2 provides a brief discussion on the established and emerging desalination processes. It also touches on the topic of renewable energy sources for desalination.

Chapter 3 presents the structure of the marine environment (i.e., seawater composition, marine habitats, and marine organisms) essential for the prediction and understanding of marine environmental impact of desalination. In addition, the chapter discusses and illustrates with real-life examples how environmental factors such as seawater quality and presence of organisms may affect the efficiency and reliability of desalination operations.

Chapter 4 describes the potential impacts of seawater desalination on the marine environment and mitigation measures. As the impacts stem from the intake of seawater and marine discharge of brine, intake and discharge systems and methods are described as well.

In several respects Chapters 5 and 6 constitute the heart of this book. They provide an extensive account on observed and reported impacts of seawater desalination on the marine environment. Chapter 5 describes early observations that range from 1960 to 2000, and Chapter 6 covers observations from 2001 to 2017. These accounts include in situ studies, in situ experiments, laboratory bioassays, and toxicity tests. The findings are critically reviewed, and generalized when possible. Shortcomings related to knowledge gaps and missing information are stressed throughout.

Chapter 7 touches on a wide scope of topics, all extensively described in the literature, which include: international and regional convention and agreements, local legislation and regulations, and the typical components of an environmental impact assessment process. Such components are hydrodynamic, biogeochemical and ecosystem based modeling, risk assessment and decision making models, toxicity testing, public engagement, and environmental monitoring.

ACKNOWLEDGMENTS

This book was written while on sabbatical from the Israel Oceanographic and Limnological Research (IOLR, Haifa) at the Monterey Bay Aquarium Research Institute (MBARI), Moss Landing, California, and at the Technion—Israel Institute of Technology, Faculty of Civil and Environmental Engineering, Haifa, Israel. Many thanks to Dr. Peter Brewer (MBARI) and Dr. Eran Fridler (Technion) for giving me the opportunity to spend my sabbatical with them working on this book.

I am grateful to Mr. Avner Hermoni (CEO) and Eng. Ofer Fine (COO) of Via Maris Desalination Ltd., Israel, for sharing their experiences on seawater desalination and the way it is affected by seawater quality. I thank them also for the permission to use two pictures in this book.

Special thanks to Ms. Hana Bernard (IOLR) for drawing Figs. 3.2, 3.3, 4.1, 4.2, and 7.2 and helping with others. Special thanks also to Ms. Michelle FIsher, my editorial project manager, for taking me through the publishing process.

Last but not least, I would like to acknowledge and thank my colleagues at the Marine Chemistry Department at IOLR, Dr. Efrat Shoham-Frider and Mr. Yaron Gertner, as well as my colleague Dr. Bella Galil for many years of joint work at sea and on land. It has been a joyous and fruitful experience.

GLOSSARY

This glossary starts with an explanation of the six most relevant topics for this book. It is followed by an expanded definition of terms used in the book and a list of abbreviations.

SALINITY (S)

Salinity is generally defined as the amount of salt (in grams) dissolved in one kilogram of seawater. However, the concept of salinity has evolved from the late 1800s and the work of M. Knudsen on the relationship between salinity and the concentration of chlorides in seawater through the 1980 equation of state (EOS-80) to the new thermodynamic equation of state (TEOS-10) adopted in 2010.[1] For the purposes of this book, we use EOS-80. Salinity is defined and measured as the ratio between the conductivity of a seawater sample and that of a standard KCl solution and, therefore, is expressed as a number, without units. However, several units can be found in the desalination literature: per mille (‰), parts per thousands (ppt, g/kg), practical salinity units (PSU), practical salinity scale (PSS). For the purposes of this book, all units express essentially an equal value.

WATER TYPES BASED ON THE AMOUNT OF DISSOLVED SOLIDS

There are no unambiguous definitions of the types of water in the literature. The following are some of the most commonly found:

- Drinking water: <500 mg/L dissolved solids. In some places defined as <1000 mg/L
- Fresh water: <1000 mg/L
- Riverine water: 500–1500 mg/L
- Brackish water: 1000–10,000 (Salinity from 1 to 10)
- Salt water: <10,000 mg/L
- Seawater: 30,000–50,000 mg/L (Salinity from 30 to 50. Typical seawater has a salinity of 35)
- Briny water: >50,000 mg/L (Salinity >50)

[1] Millero, F.J., 2010. History of the Equation of State of Seawater. Oceanography 23, 18–33.

UNITS FOR PLANT CAPACITY

- 1 cubic meter (m^3) = 264 gallons = 0.0008 acre-feet (AF)
- In this book, plant capacity is given as m^3/day (m^3/d)

CHARACTERIZATION OF DESALINATION PLANTS BY CAPACITY

- Small: $<1000\,m^3/d$
- Medium: 1000–10,000 m^3/d
- Large: 10,000–50,000 m^3/d
- Extra large: $>50,000\,m^3/d$
- Mega: $>100,000\,m^3/d$

INSTALLED CAPACITY VS ACTUAL PRODUCTION

Installed capacity is the maximal amount of desalinated water a plant can produce. The actual production is either the installed capacity or less. To understand the possible effects of seawater desalination on the marine environment, it is essential to know and address the actual production and not the installed capacity of a desalination plant.

TERMINOLOGY FOR FEEDWATER, BRINE, AND PRODUCT WATER

- Feedwater is the stream of water fed into the desalination process. Also found in the literature as: input water, raw water, source water.
- Brine is the stream of liquid discharged from a desalination plant. Also found in the literature as: concentrate, retentate, effluent, discharge, reject stream, concentrated salt stream. In this book, brine includes the concentrated seawater and the chemicals used in the desalination process discharged jointly at the marine environment.
- Product water is the water produced as a result of the desalination process. Also found in the literature as: permeate, condensate, finish water, potable water, distillate.

DEFINITION OF TERMS

Abyssal plain	The area encompassing the ocean floor where the water depth is deeper than 4 km
Anion	A negatively charged atom or molecule
Anoxia	Total depletion of dissolved oxygen in water
Anthropogenic	Originating in human activity
Antifoaming	Compounds used in thermal desalination processes to prevent foam formation and carryover of brine into the distillate
Antiscalants	Compounds used in the desalination process to prevent the deposition of inorganic salts
Aphotic zone	The dark region of the ocean that lies below the sunlit upper layer water
Archea	Single cell, prokaryotic (without a cell nucleus) microorganisms. They may be autotrophic, heterotrophic or decomposers
Autotrophs	Organisms that synthesize their food from inorganic compounds. Marine autotrophs produce organic matter from dissolved components in seawater
Bacterioplankton	Single cell, prokaryotic (without a cell nucleus) microorganisms. They may be autotrophic, heterotrophic or decomposers
Bar	Unit of pressure. Equals 105 Pascals and 0.9869 Atmospheres
Benthic	The zone of the ocean adjoining the sea bed
Benthos or benthic organisms	Bottom dwelling marine organisms. They live on the ocean floor, either on the substrate (epifauna and epiflora) or buried or burrowing in the sediment (infauna)
Biofouling	The adhesion of microorganisms on a membrane (or along a water treatment system) to form a biofilm that can further grow and entrap organic molecules, particles and other microbial cells. It reduces the system's performance
Biogenic	Produced by living organisms
Brine	A very saline solution. In desalination it refers to the plants discharge
Cation	A positively charged atom or molecule

Continued

Coagulants	Compounds added to the feedwater to cause small charged particles to agglomerate and form larger particles that are easier to remove
Commons	Area where resources are available to all claimants, not subject to the control of any person or government such as the open ocean
Conductivity	The ability of a solution to conduct an electrical current
Continental margin	The offshore zone, consisting of the continental shelf, slope, and rise. It separates the dry-land portion of a continent from the deep ocean floor
Continental rise	The sea floor following the continental slope. It continues to deepen but with a slope gentler than the continental slope
Continental shelf	The shallowest part of the continental margin. Extends seawards at a gentle bottom slope from the coast up to 120–200 m depth. Almost all seawater desalination plants draw feedwater from and discharge brine at the continental shelf
Continental slope	The ocean's bottom having a steep slope. It extends from the shelf break down to 2–3 km depth
Decomposer	Organisms that break down detritus for nutrition
Density	Mass per unit volume
Desalination	Process that removes dissolved solids, primarily salts and other constituents from a saline water source. Also referred to as desalinization and desalting
Detritus	The particulate, organic remains and waste of organisms. Constitutes a major food source in marine ecosystems
Dissolved oxygen saturation	The maximal amount of oxygen that can be dissolved in water at a specific temperature and salinity (100% saturation)
Distillation	The purification of liquids by boiling. Distilled vapor is collected and condensed into a pure liquid
Ecosystem	The combined habitat, its living organisms and energy source
Ecosystem approach	A strategy for the integrated management of land, water and living resources that promotes conservation and sustainable use in an equitable way
Electrodialysis	The separation of a solution's ionic components through the use of semipermeable, ion-selective membranes, operating in an electric field

Emission limits	Numerical values (or narrative statements) for the concentrations of effluent components at the point of discharge. Also known as effluent standards or discharge quality standards
Entrainment	The transport of organisms by the flow of seawater through the intake's screens and into the desalination plant, removing them from the marine environment
Entrapment	The trapping of organisms within the intake system. They remain alive and can thrive but cannot escape back to their natural habitat
Environmental quality standard	The concentrations of constituents permitted beyond the mixing zone. Also known as ambient or reference standards
Epipelagic	The upper region of the ocean extending to a depth of about 200 m
Euphotic or photic zone	Near surface layer of water of the ocean where sunlight penetrates enough for photosynthesis to occur
Far field	The region where mixing and dilution of a discharge are caused by oceanic turbulence, currents and stratification
Feedwater	Input or raw water stream fed into the desalination process
Flocculants	Compounds used to agglomerate noncharged particles, in the feedwater, into larger particles that can be removed by sedimentation or flotation
Foaming	Occurs in thermal desalination processes when organic compounds are concentrated in the brine, giving rise to surface active effects, which increase the liquid film strength at phase interfaces
Food web	Food and energy flow in an ecosystem (production, consumption, and decomposition) and the organisms involved in these processes
Fossil water	Groundwater created during ancient climate conditions, not renewable under current conditions. Also known as paleowater
Fouling	The reduction in performance of a process that occurs as a result of scaling, biological growth, or deposition of colloidal matter
Groundwater	Water found below the Earth's surface in geological reservoirs known as aquifers
Gulf or The Gulf	Arabian or Persian Gulf
Habitat	The natural home or the physical environment in which an organism lives
Haline	Relating to the degree of saltiness or salinity

Continued

Harmful algal bloom or Red tide	A rapid increase in the abundance of phytoplankton or benthic algae that may harm the environment. Some bloom species taint the water red, thus the reference as Red tide
Heterotrophs	Organisms that cannot manufacture their own food and therefore derive nutrition mainly from plant or animal matter
Hybrid system	Integration of different combinations of desalination processes as well as dual purpose technologies used to coproduce energy and fresh water
Hydrological cycle	The Sun-driven process of evaporation, condensation, and precipitation that circulates water from the oceans and Earth to the atmosphere and back. Also known as the water cycle
Hypoxia	Low (<30%) saturation of dissolved oxygen in water
Impingement	The pinning of organisms against the open intake's screens by the velocity and force of the feedwater flowing through them into the desalination plant
In situ	From Latin, in its original place. In oceanography, used when measuring properties directly in the sea, without sampling
Infauna	Animals that live in soft sediments
Intake systems	Structures used to extract feedwater and convey it to the desalination plant
Intertidal zone	The littoral zone subjected to the influence of tides
Invasive alien species	Nonindigenous organisms, from all taxonomic groups, introduced accidentally or deliberately by human activities into an environment where they are not normally found
Larvae	Immature life stage or form of an animal
Major constituents	Compounds in seawater with concentration higher than 100 mg/kg
Membrane	A semipermeable film, in the context of desalination
Membrane process	A nonphase change desalination process, in which semipermeable membranes are used to separate product water from dissolved salts
Microfiltration	Filtration through 0.03–10 μm. Molecular weight cut off of >100,000 Daltons
Minor constituents	Compounds in seawater with concentration from 1 to 100 mg/kg

Mixing zone	An allocated impact zone where environmental quality standards can be exceeded as long as acutely toxic conditions are prevented. It is a regulatory concept
Monitoring	The systematic, repeated measurement of biotic and abiotic parameters of the environment with a predefined spatial and temporal design
Multieffect distillation	A thin film evaporation process where the vapor formed in a chamber, or effect, condenses in the next, providing a heat source for further evaporation
Multistage flash evaporation	A desalination process where a stream of feedwater flows through the bottom of chambers, or stages, each operating at a successively lower pressure. A proportion of it flashes into steam and is then condensed
Nanofiltration	Filtration through 0.001 μm–1 nm. Molecular weight cut off of <1000 Daltons
Near field	The region where the mixing of a discharge in the environment is caused by the turbulence generated by the discharge itself
Nekton	Living organisms that are able to swim and move independently of currents
Neritic	The water that overlies the continental shelf, generally shallower than 200 m
Neritic pelagic zone	The water mass located above the continental shelf, almost all within the euphotic zone
Nonpoint source	Multiple, not easily identifiable, sources of pollution also called diffuse source
Nutrients	Inorganic or organic compounds necessary for the nutrition of autotrophs
Oceanic waters	The waters beyond the shelf break, generally at water depth greater than 200 m
Organic fouling	Accumulation of dissolved, colloids, and particulate organics, such as humic substances, on the surface or in the membrane pores
Osmoregulation	The active process by which an organism maintains its level of water and hence its salt content
Osmosis	The natural process by which water passes across a semipermeable membrane from an area of higher water concentration (more dilute) to one of lower water concentration (more saline)
Osmotic pressure	Excess pressure that must be applied to a concentrated solution to cancel out osmosis

Continued

Outfall system	The infrastructure through which effluent is discharged into a receiving water body
Particulate matter	Particles with a diameter larger than 0.2 μm, although a cut off of 0.45 μm is also used. Particles may be suspended in water
Pelagic	The domain of all water in the oceans
Per mille	Parts per thousand
Permeate	The portion of the feedwater that passes through a membrane
Photosynthesis	The synthesis of organic material using carbon dioxide, water, inorganic salts, and light energy (usually sunlight) captured by light-absorbing pigments, such as chlorophyll
Phytoplankton	Single cell, eukaryotic (contain a nucleus), autotrophic microorganisms
Plankton	Organisms that drift with the currents. They may have limited self-movement
Point source	A single, identifiable, source of pollution
Posttreatment	Partial remineralization of the product water prior to distribution
Pretreatment	Processes to remove suspended particles and colloids from the feedwater
Primary productivity	The rate at which photosynthetic organisms produce organic compounds in an ecosystem
Red tide	See harmful algal bloom
Reverse osmosis (RO)	A method of separating water from dissolved salts by passing feedwater through a semipermeable membrane at a pressure greater than the osmotic pressure
Scaling	Inorganic fouling, caused by the precipitation of inorganic salts, such as calcium carbonate, calcium sulfate, barium sulfate, and silicates
Sea shore	The area where land meets the sea
Sessile	Organisms that live attached to a substrate or surface, not free to move about
Shelf brake	The part of the ocean floor identified by an abrupt increase on the steepness of bottom slope. It is considered as the boundary between the coastal and oceanic habitats
Stratification	The layering of water masses based on density. The lighter water mass is the upper layer. Density increases with increasing depth

Thermal process	A desalination process involving a phase change. Feedwater is heated under suitable operating temperatures and pressures and the vapor formed is condensed as pure water
Thermocline	A distinct layer in a large water body in which temperature changes more rapidly with depth than it does in the layers above or below it
Trace constituents	Compounds in seawater with concentration lower than 1 mg/kg
Trophic level	Stage in the food web. Usually determined by feeding relationship among organisms
Turbidity	The measure of the clarity of a liquid, expressed as the amount of light scattered by material in the water. The more scattering, the higher the turbidity
Ultrafiltration	Filtration through 0.002–0.1 μm. Molecular weight cut off of 10,000–100,000 daltons
Water mass	A water body usually characterized by its temperature and salinity and by some other tracers
Water-energy-food nexus	The integrated management of natural resources to ensure water, energy, and food security in a sustainable way
Withdrawal	Water removed from a source and used to meet a human need
Zooplankton	Single or multicellular heterotrophic organism

LIST OF ABBREVIATIONS

2D	two-dimensional
3D	three-dimensional
AD	adsorption desalination
AHP	analytic hierarchy process
ANZECC	Regional Australian and New Zealand Environment Conservation Council
AS	antiscalant
ASTM	American Society for Testing and Materials
BACI	before and after and control and impact
BAT	best available technology
BEP	best environmental practice
BOO	build, operate, own
BOOT	build, own, operate, transfer

Continued

BOT	build, operate, transfer
BTEX	benzene, toluene, ethylbenzene, and xylene
BW	brackish water
CA	California
CAPEX	capital expenditure
CBD	convention on biological diversity
Chl-*a*	chlorophyll-*a*
CHs	clathrate hydrates
CIP	clean in place
d	day
DO	dissolved oxygen
DOM	dissolved organic matter
DPSIR	driver-pressure-state-impact-response
EC_{50}	concentration that causes an effect on 50% of the population
EcAp	ecosystem approach
ED	electrodialysis
EI&E	entrainment, impingement and entrapment
EIA	environmental impact assessment
ELV	emission limit value
EPA	Environmental Protection Agency
EQS	environmental quality standards
ESA	European Space Agency
EU	European Union
FAO	Food and Agriculture Organization
FD	freeze desalination
FO	forward osmosis
GES	good environmental status
GHG	greenhouse gas
GIS	Geographic Information System
GWI	Global Water Intelligence
HAB	harmful algal bloom
HDH	humidification dehumidification
IAS	invasive alien species
IC_{50}	concentration that causes an inhibition of growth of 50% in unicellular algae bioassay
ICM	Integrated Coastal Management
ICP	ion concentration polarization
ICZM	Integrated Coastal Zone Management
IDA	International Desalination Association
IOC	Intergovernmental Oceanographic Commission
ISO	International Organization for Standardization

IUCN	International Union for the Conservation of Nature
KAUST	King Abdullah University of Science and Technology
KSA	Kingdom of Saudi Arabia
LOEC	lowest observed effect concentration
m^3	cubic meter
MAP	Mediterranean Action Plan
MCDA	multicriteria decision analysis
MCDI	membrane capacitive deionization
MD	membrane distillation
MDC	microbial desalination cell
MED	multieffect distillation
MENA	Middle East-North Africa
Mm^3	million cubic meters
MPA	Marine Protected Areas
MSF	multistage flash
MSFD	Marine Strategy Framework Directive
NASA	National Aeronautics and Space Administration
NF	nanofiltration
NOAA	National Oceanic and Atmospheric Administration
NOEC	no observed effect concentration
NO_x	nitrate and nitrite
NR	not reported
NRC	National Research Council
NTU	nephelometric turbidity unit
OECD	Organization for Economic Co-operation and Development
OPEX	operating expenditure
PERSGA	Regional Organization for the Conservation of the Environment of the Red Sea and Gulf of Aden
PO_4	phosphate
ppb	parts per billion
ppm	parts per million
ppt	parts per trillion
PRO	pressure retarded osmosis
PV	pervaporation
QA	quality assurance
QC	quality control
RE	Renewable energy
RED	reverse electrodialysis
RO	reverse osmosis
ROPME	Regional Organization for the Protection of the Marine Environment of the Gulf States and Oman

Continued

S	salinity
Sat	saturation
SDG	sustainable development goal
SDI	silt density index
SEA	strategic environmental assessment
SPM	suspended particulate matter
SW	seawater
SWRO	seawater reverse osmosis
T	temperature
TDS	total dissolved solids
TEP	transparent exopolymer particle
TOP	total organic phosphorus
UAE	United Arab Emirates
UF	ultra-filtration
UN	United Nations
UNEP	United Nations Environment Program
UNESCO	United Nations Educational, Scientific and Cultural Organization
USA, US	United States of America
WEF	Water-energy-food
WET	Whole Effluent Toxicity
WFD	Water Framework Directive
WHO	World Health Organization
ZID	zone of initial dilution
ZLD	zero liquid discharge

CHAPTER 1
Introduction

How inappropriate to call this planet "Earth," when it is clearly "Ocean"
Arthur C. Clarke

I chose to start this book with the ocean rather than with the obvious and more commonplace introductory statements like: "the growing global population", "the increase in freshwater demand", "the dwindling of the natural water sources" followed by "desalination as a solution". This book is centered on the oceans as the source for freshwater supply through desalination and on the impact of desalination on the marine environment. However, the oceans are much more than a supplier of saline water to the desalination industry. They regulate climate and are vital for the preservation of life on Earth.

Most of the water (96.5%) on Earth is saline, undrinkable, and held within the oceans. Fresh water comprises 2.5% of the total water, with 1.7% in glaciers and ice caps, and thus inaccessible for use. Only 0.8% of the total global water is fresh and can be withdrawn for consumption. Moreover, the global distribution of fresh water is uneven; some areas are rich in water resources, such as the Amazon rainforest, while others are arid such as the Middle-East—North Africa region (Fig. 1.1). To overcome this uneven distribution, the ancient Romans transported water through aqueducts, Persian engineers constructed qanats (tunnels) to carry groundwater from high elevations down to dry areas, and, in 1991, Libya constructed the great man-made river that brought fossil groundwater from the Sahara to the coast.

Today, this uneven distribution of fresh water is exacerbated by population growth, increased use of water per capita, and climate change. Climate change affects weather and precipitation patterns and decreases the reliability of natural freshwater supply. Desalination is thus placed as a viable and reliable new source of water. Seawater desalination is especially appealing as half of the world's population and 75% of the large cities (>5 million inhabitants) are located within 100 km from the coast. Desalination is recommended by

Marine Impacts of Seawater Desalination: Science, Management, and Policy
https://doi.org/10.1016/B978-0-12-811953-2.00001-3

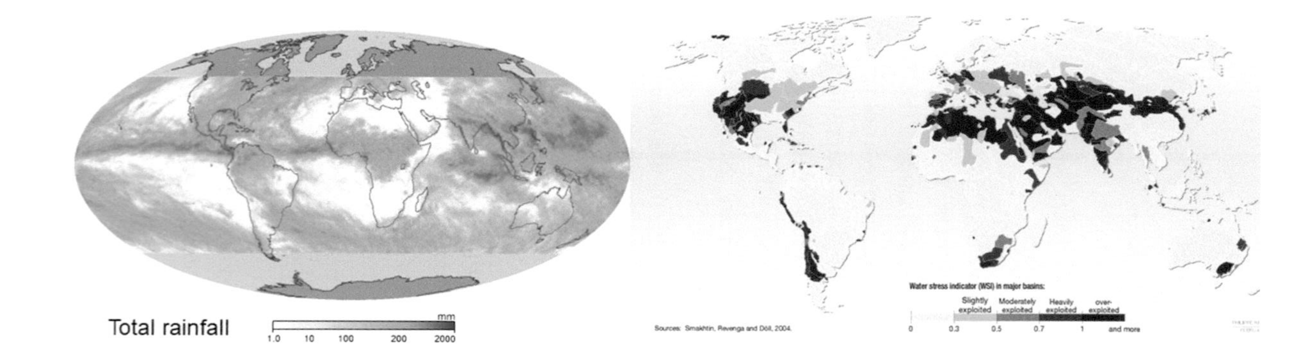

Fig. 1.1 Global freshwater distribution. The uneven distribution is depicted in the total monthly rainfall in millimeters as recorded by NASA's Tropical Rainfall Measuring Mission on August 2016 (left panel) and the map of water scarcity indicator (WSI) (right panel). *(Reproduced with permission from NASA Earth Observatory (left panel) and from GRID-Arendal (right panel) at http://www.grida.no/resources/5586, Philippe Rekacewicz, February 2006.)*

the United Nations, through the goals of Agenda 2030 for Sustainable Development, as an essential tool to provide clean water and sanitation to the world's population. However, desalination is expensive. The capital expenditure to build a large seawater desalination plant ranges from $200 to $600 million, and the operating expenditure ranges from $0.5/m^3 to $2/m^3. Both expenses are highly dependent on the plant's individual features and on national policy.

This introductory chapter provides a concise overview on topics associated with seawater desalination that are not otherwise covered in this book: the early history of seawater desalination, the desalination market, water policy, and the quality of desalinated water.

1.1 A BRIEF HISTORY OF SEAWATER DESALINATION

Distillation is one of mankind's earliest ways of obtaining fresh water from seawater. It mimics the hydrological cycle: when salt water is boiled, the heat causes water to evaporate, leaving the salt beyond. The vapor is cooled, recondensed and the fresh water is collected. Desalination and distillation are mentioned in biblical texts and in ancient Greek and Roman writings. In the Book of Exodus, Moses transformed the bitter waters at Marah to sweet water using a "tree"—a kind of purification method using plants—and the hydrologic cycle or distillation are mentioned in the Book of Job. Aristotle (384–322 BC) wrote: "Experiment has taught us that seawater when converted into vapor becomes potable, and the vaporized product, when condensed, no longer resembles sea-water." Pliny the Elder (AD 23–79) describes a method of hanging fleece over the side of a ship at night, just above the surface of the water, to collect water vapor during the evening and squeeze it out in the morning to provide fresh water.

The art of distillation advanced between the 1st and 3rd centuries AD with Maria the Jewess, the Greek Egyptian alchemist Cleopatra (both in Alexandria), and Alexander of Afrodisia. Their effort was geared mainly toward the production of scented oils and perfumes from natural products. This technology led to the development of the alembic condenser (Fig. 1.2) described by Zosimos of Panopolis. A process of boiling seawater over a fire on board a ship, with a natural sponge placed over the mouth of the container, was described in the 4th century AD. The water vapor condensed in the sponge, and water was squeezed out for drinking. The knowledge on distillation and work of early Greek, Persian, Egyptian and Islamic scholars was brought to Western Europe with the Moorish conquest.

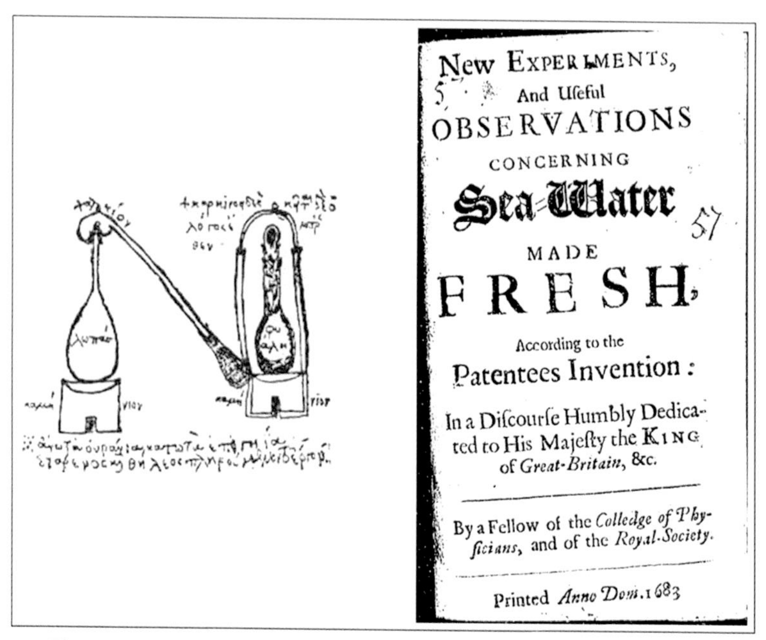

New EXPERIMENTS,
And Uſeful
OBSERVATIONS
CONCERNING
Sea-Water
MADE
FRESH,
According to the
Patentees Invention:
In a Diſcourſe Humbly Dedica-
ted to His Majeſty the KING
of *Great-Britain*, &c.

By a Fellow of the *Colledge of Phy-ſicians*, and of the *Royal-Society*.

Printed *Anno Dom.* 1683

Fig. 1.2 The Zosimos distillation apparatus, called Alembic, depicted by an unknown Byzantine Greek illustrator and reproduced by Marcelin Berthelot in 1887 (left panel, Wikimedia Commons). In the right panel is a picture of the first page of the publication by Nehemiah Grew (1641–1711) describing seawater desalination based on patents submitted by others.

Leonardo da Vinci (1452–1519) accurately described the hydrological cycle and suggested that pure water can be produced from a simple still on a kitchen stove. Seawater desalination to supply fresh water for sailors expanded from the late 1500s to 1700, the age of sea voyages. In particular, during the reign of Charles II (1660–85), patents for seawater desalination were issued in England (Fig. 1.2). Stephen Hales (1677–1761) discovered that the first product of the distillation (one-third of the seawater volume) was of much higher quality than the rest. Thomas Jefferson (1743–1826) published the "Report on the method for obtaining fresh water from salt" in 1791 stating, "And it would seem that all mankind might have observed that the earth is supplied with fresh water chiefly by exhalation from the sea, which is in fact an insensible distillation effected by the heat of the sun."

In 1748 J.A. Nollet (1700–1770) discovered the process of osmosis in which a solvent will pass spontaneously through a membrane, in his case an animal bladder, from a dilute solution into a more concentrated one. By applying pressure to the more concentrated side, the flow of solvent could be slowed, stopped, or reversed, which led to the term reverse osmosis (RO). About two hundred years after this discovery, the process of reverse osmosis was developed for seawater desalination (see Chapter 2).

No real improvement of the distillation process occurred until the 1800s with the advent of steam engines and the need for pure water for marine boilers and trains. A patent was issued in 1897 for a mechanically driven vapor compression system and in 1900 for the concept of multi-stage flash distillation (MSF) (see Chapter 2). Distillers were installed in 1907 in Jeddah (KSA, Red Sea), in 1910 near Safaga, and in 1933 in Qusair (Egypt, Red Sea). A thermo-compression evaporator was installed in 1928 in Curacao. Membrane separation and electrodyalysis appeared between 1920 and 1930 but were not commercial because of lack of appropriate semipermeable membranes.

The first solar desalter was installed in 1872 in Las Salinas, Chile, with a capacity of 19 m^3/day. Seawater distillation powered by solar energy was mentioned in 1943 and used in WWII in life rafts on ships and in 1953 as a possible water supply process in tropical islands. In addition, a chemical process was used to produce a compact, sealed, desalination kit for trans-ocean fliers forced down at sea during WWII. It was based on the addition of a base to precipitate the anions present in seawater and of acid to precipitate the cations. Each step was followed by filtration of the precipitate formed. The final product was "not unpleasant in taste."

Desalination on a commercial scale started in 1957 using thermal processes with the onset of flash evaporation and MSF processes. The use of membrane processes (mainly RO) began to grow following the development of the commercial spiral wound membranes by Loeb and Sourirajan in 1963 and has been increasing since.

1.2 THE WATER DESALINATION MARKET

The global water desalination market, which has been expanding since the onset of commercial desalination, was evaluated at $13 billion in 2016 and is expected to double its worth by 2025.

Data on the desalination market are compiled and managed mainly by the International Desalination Association (IDA) and the Global Water Intelligence (GWI). IDA, a nonprofit organization, promotes research and information dissemination to the community and to the public. GWI is geared toward industry, investors and stakeholders in water management. It tracks major water projects and business opportunities and provides market forecasts.

Most statistics on desalination originate from these two sources. By the mid-1960s there were only 26 desalination plants with a capacity larger than 3500 m^3/day. All but one used thermal processes and 24 desalinated seawater. From 1997 to 2008 the compound annual growth rate of desalination was 17% with 58 Mm^3/day installed capacity in 2008. Desalination grew exponentially at a rate of 14% per year from 2007 to 2012, when it reached 74.8 Mm^3/day. From 2012 to 2015 the increase rate declined to 3% per year. By October 2017 there were 19,372 desalination plants in 150 countries with 99.8 Mm^3/day contracted capacity and 92.5 Mm^3/day with actual production capacity. About 65% of the capacity was used for seawater desalination, and the remaining capacity was used for brackish and inland waters desalination.

1.3 THE WATER-ENERGY-FOOD-ECOSYSTEM NEXUS

Access to fresh water and energy is critical to human well-being and to socio-economic development. Water and energy are intertwined: water is needed to mine and extract energy-producing resources, such as coal, uranium, oil, and gas, and to irrigate crops for biofuel production, such as corn and sugar cane. Water is also used for cooling power plants and to drive hydroelectric and steam turbines. Similarly, energy is needed to withdraw water from its source and transport it to the end users. Most importantly, energy is the main resource needed for desalination.

The demand for energy and water are increasing globally. Seawater desalination provides a new source of water, some say almost infinite, in contrast to the finite supply of fresh water from nature. Concurrently, the demand for food is increasing as well. Food production, through agriculture, is the largest user of water. It uses 70% of the global water withdrawal and consumes one-third of the energy production through its supply chain. Industry and households account for the remaining 20% and 10% of the water use, respectively. The concept of invisible (or virtual) water is now widespread and refers to the transfer of water not in its liquid state but as food and commodities. For example, it takes about 2500 L of water to produce 3 kg of alfalfa for a cow to produce 3.8 L of milk, or to produce one cotton shirt.

These three sectors—water, energy, and food—are inextricably linked. The actions taken in one area may impact the others. This linkage gave way to the water-energy-food (WEF) nexus approach, the integrated management of natural resources to ensure water, energy, and food security in a sustainable way. The WEF nexus approach also strives to reconcile the use of natural resources with the protection of the environment and ecosystem functions, adding them to the nexus as an important element. Allocation of water to the environment is critical for its health, while a damaged ecosystem can in turn affect the quality of water. The nexus approach can help prevent the effects of climate change, already impacting the water cycle and the environment, from cascading down to water, energy, and food availability.

1.4 NATIONAL AND INTERNATIONAL WATER POLICY

The establishment of seawater desalination as an economically viable and reliable fresh water source changed national and international attitudes toward water and its management. Desalination removes, to a certain extent, the quantitative restrictions on water supply and allows for a better control of its quality. In particular, desalination increases drought resilience in the presence of climate change.

At the national level, water is a crucial resource and a public good. The supply of water is managed and controlled by the state; it produces the water and transports it from natural sources (ground water, rivers, lakes) and man-made ones (dams and reservoirs) to the end users. Implementation of sea-water desalination turns water into a manufactured good, a commodity. In many cases the states delegate these activities to the private sector, mostly multinational partnerships that can handle such large-scale projects. Usually, the private water production is supervised by the state that is also committed to purchase the produced water. Financing schemes to build and operate a desalination plant include: BOOT (build, own, operate, transfer), BOT (build, operate, transfer), and BOO (build, operate, own).

At the international level, the trans-boundary management of water, or "hydro-diplomacy," consists of joint supervision of water quality, collaboratively effort for mitigating floods, hydropower production, and ecological management of shared resources. Prior to the era of significant seawater desalination there was an inherent advantage to nations being "upstream" near the water source. Seawater desalination shifts this advantage to "downstream" nations with coastal access that are now able to control their national water supply. Coastal nations can even export the produced water upstream or to neighboring countries either within the framework of international treaties or by selling it as a commodity.

1.5 DESALINATED WATER QUALITY

Water produced by desalination is low in minerals and slightly acidic and has low buffering capacity. Used as such, without posttreatment, it may corrode metals and dissolve concrete within its storage and distribution systems. In the posttreatment, the product water is remineralized by adding chemical constituents as calcium and magnesium carbonate (such as percolation through limestone) along with pH adjustment or through blending with small volumes of mineral-rich waters (see Section 2.2.1.2).

The lack of essential minerals, such as calcium and magnesium ions, has the potential to affect public health and agricultural practices in places where desalinated water is the main source of fresh water. Although drinking water typically contributes a small proportion to the recommended daily intake of essential elements, it could be important in some regions, such as in the Gulf states (KSA, UAE for example) and Israel.

In contrast to the low mineral content of desalinated water, some constituents remaining in the product water may affect human health and crops. For example, RO has a low rejection for boron, which has a natural concentration of 4.5 mg/L in seawater (see Chapter 3). Health based guideline for boron in drinking water is 2.4 mg/L while some crops are sensitive at levels of 0.5 mg/L. Therefore, a second RO pass is implemented to achieve these limits. In regions with substantial oil extraction activity, the product water may be contaminated by volatile organics, such as benzene, toluene, ethylbenzene, and xylenes. These volatile organics have health-based drinking water guidelines and can cause unacceptable taste and odor at lower concentrations. Although thermal processes are designed to vent those gases during the desalination process, they still need to be monitored. High concentrations of these volatiles can potentially dissolve RO membranes and thus appear in the product water.

1.6 THE WAY FORWARD

A detailed outlook on seawater desalination, its cost to the marine environment, and mitigation measures for reducing the environmental impact is integrated within the book's chapters. However, it is fitting to conclude this introduction with a short synthesis of desalination's environmental implications into the future.

First and foremost, seawater desalination is expensive both financially and environmentally. Moreover, many countries cannot afford its cost, thus excluding desalination as an option to increase freshwater supply. Therefore, basic measures to increase water availability should be implemented before resorting to desalination. Such measures are: reduced consumption, prevention of water loss by amelioration of infrastructures, capture and storage of rain and storm water, treatment and recycling of used water, and reuse in agriculture and industry. The theme of the world's water day in 2017, sponsored by the United Nations, was "Why waste water?" aiming to take action to tackle the water crisis.

When the option to desalinate seawater is chosen, it should be managed at two different but interconnected levels: (1) technological research and development, and (2) environmental management and regulation. Desalination can be improved by increasing yield and selectivity, reducing energy consumption, and reducing waste discharge and the use of chemicals (see Chapter 2). These aims can be achieved by several methods: improvement of existing membranes and development of new materials; utilization of renewable energy or heat waste in existing processes; development of hybrid systems able to efficiently combine energy and water production; reaching zero liquid discharge and water and salts recovery; improvement of existing technologies; use of less chemicals in the process and more environmental friendly compounds; and development novel desalination technologies. At the environmental management and regulation aspects, seawater desalination should be an integral part of the wider coastal zone management and of marine spatial planning, addressing also trans-boundary transport (see Chapter 7). A global network for data and knowledge exchange and dissemination on the environmental impact of seawater should be established. This future network is envisaged to provide an essential tool for regulators to develop guidelines and laws to protect the environment. This data network can be a source of knowledge to stakeholders providing the tools to understand and evaluate the effects of a desalination project.

The oceans, an integral part in freshwater production through seawater desalination, should be managed with respect and in a sustainable way for the well-being of Earth and the generations to come.

REFERENCES

Loeb, S., Sourirajan, S., 1963. Sea Water Demineralization by Means of an Osmotic Membrane, Saline Water Conversion—II. American Chemical Society, pp. 117–132.

FURTHER READING

Aviram, R., Katz, D., Shmueli, D., 2014. Desalination as a game-changer in transboundary hydro-politics. Water Policy 16, 609–624.

Birkett, J.D., 1984. A brief illustrated history of desalination. Desalination 50, 17–52.

Birnhack, L., Voutchkov, N., Lahav, O., 2011. Fundamental chemistry and engineering aspects of post-treatment processes for desalinated water—a review. Desalination 273, 6–22.

Burn, S., Hoang, M., Zarzo, D., Olewniak, F., Campos, E., Bolto, B., Barron, O., 2015. Desalination techniques—a review of the opportunities for desalination in agriculture. Desalination 364, 2–16.

Desalination Expert Group, 2014. Desalination in the GCC. The History, the Present & the Future. The Cooperation Council for the Arab States of the Gulf (GCC).

earthobservatorynasa.gov.

Gude, V.G., 2016. Desalination and sustainability—an appraisal and current perspective. Water Res. 89, 87–106.

idadesal.org.

Leighton, D., Nusbaum, I., Mulford, S., 1967. Effects of waste discharge from point Loma saline water conversion plant on intertidal marine life. J. Water Pollut. Control Fed. 39, 1190–1202.

Lior, N., 2017. Sustainability as the quantitative norm for water desalination impacts. Desalination 401, 99–111.

March, H., 2015. The politics, geography, and economics of desalination: A critical review. Wiley Interdiscip. Rev. Water 2, 231–243.

Office Saline Water, 1968. A study of the discharge of the effluent of a large desalination plant. Research and Development, Progress Report No. 316.

Sowers, J., Vengosh, A., Weinthal, E., 2011. Climate change, water resources, and the politics of adaptation in the Middle East and North Africa. Climate Change 104, 599–627.

Spealman, C.R., 1944. The chemical removal of salts from sea water to produce potable water. Science 99, 184–185.

Spiritos, E., Lipchin, C., 2013. Desalination in Israel. In: Becker, N. (Ed.), Water Policy in Israel: Context, Issues and Options. Springer Netherlands, Dordrecht, pp. 101–123.

Telkes, M., 1953. Fresh water from sea water by solar distillation. Ind. Eng. Chem. 45, 1108–1114.

Tiger, H.L., Sussman, S., Lane, M., Calise, V.J., 1946. Desalting Sea water. Ind. Eng. Chem. 38, 1130–1137.

United States, 1968. The A-B-Seas of Desalting.

WHO, 2011. Safe drinking-water from desalination. World Health Organization, WHO/HSE/WSH/11.03.

World Bank, 2016. High and Dry: Climate Change, Water, and the Economy. The World Bank, Washigton, DC.

WWAP (United Nations World Water Assessment Programme), 2014. The United Nations World Water Development Report 2014: Water and Energy. UNESCO, Paris.

www.desaldata.com.

www.fao.org/nr/water/aquastat/maps/index.stm.

www.globalwaterintel.com.

www.unwater.org/water-facts/water-food-and-energy.

CHAPTER 2

Desalination Technologies

This chapter on desalination technologies is brief, aiming at introducing the reader to the established and emerging desalination processes, citing the extensive literature on the subject. It also touches on the topic of renewable energy (RE) sources for desalination. The emphasis of the chapter is on seawater (SW) desalination, although many of the processes are also used for brackish water (BW) desalination and waste water treatment. Although concise, the description of the processes sets the stage for the discussion on the possible environmental impacts associated with the desalination industry addressed in the subsequent chapters.

2.1 GENERAL CLASSIFICATION OF DESALINATION TECHNOLOGIES

Desalination is classified into two major processes based on the technology for producing fresh water: (1) membrane processes in which semipermeable membranes separate product water from dissolved salts, and (2) thermal processes in which feedwater is heated under suitable operating temperatures and pressures and the vapor condensed as pure water. Thermal processes dominated the desalination industry up to 2005, when the use of membrane processes surpassed them, due to the rapid development and improvement in membrane technology, leading to increased efficiency and reducing operational costs. Membrane processes now encompass about 65% of the global desalination effort. Hybrid systems integrate different combinations of membrane and thermal desalination processes as well as dual purpose technologies used to coproduce energy and fresh water. Novel technologies, such as microbial desalination cell (Section 2.5.3.1) are now at the early stages of development. Regardless of the process, they all share some common requirements: (1) pretreatment of the feedwater to reduce fouling, scaling, and other interferences to the process, (2) brine management, (3) product water post-treatment prior to distribution, and (4) energy consumption and water production optimization. Fig. 2.1 depicts the established and

Marine Impacts of Seawater Desalination: Science, Management, and Policy
https://doi.org/10.1016/B978-0-12-811953-2.00002-5

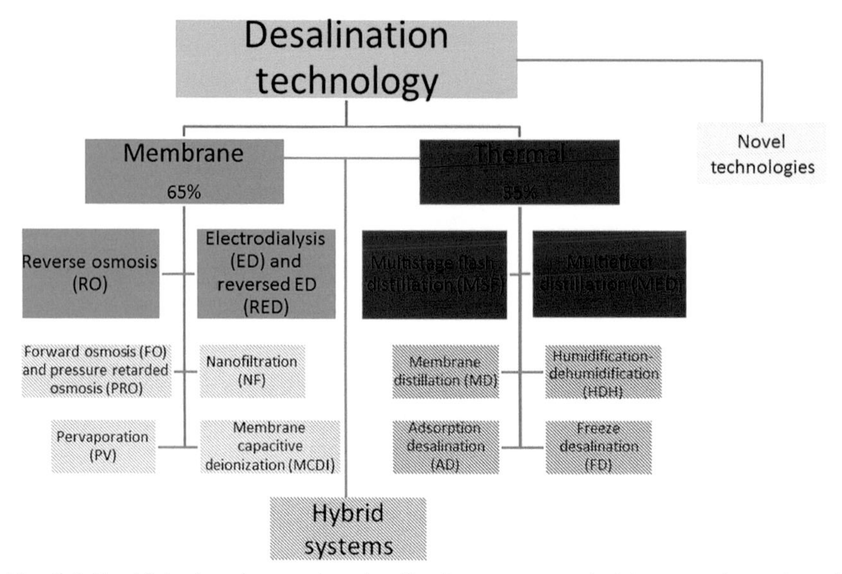

Fig. 2.1 Established and emerging desalination processes. In blue, membrane-based technologies and in red, thermal-based technologies. Membrane processes, mainly reverse osmosis, constitute 65% of the total global desalination.

emerging desalination processes, and Table 2.1 compares among their characteristics and specific features.

2.2 MEMBRANE PROCESSES

Membrane processes are nonphase change processes in which water remains in the liquid phase, and semipermeable membranes separate water or salt from the feedwater. These processes are driven by electrical power or by natural osmotic pressure gradient. Established and emerging membrane processes include: reverse osmosis (RO), electrodialysis (ED), reverse electrodialysis (RED), forward osmosis (FO), pressure retarded osmosis (PRO), ultrafiltration (UF), membrane capacitive deionization (MCDI) (see Sections 2.2.2–2.2.5).

Overall, membrane processes have common operational goals: maximization of membrane selectivity, permeability and mechanical strength, usually a trade-off among them; reduction of concentration polarization across the membranes; and reduction of membrane fouling and scaling. As mentioned, membrane processes need to: (a) pretreat the feedwater, the extent depending on feedwater quality and on the desalination process; (b) to

Table 2.1 Comparison among different desalination processes

Process	Strengths	Limitations	Outlook	References[a]
Reverse osmosis (RO)	• Most efficient process for seawater desalination • Compared to thermal processes: lower energy requirements, higher water recovery, lower capital and operating costs • Low price membranes • No thermal pollution • Commercially operating mega-sized plants with lower operating costs due to economy of scales • Adjustable production based on demand	• External pressure application required • Maximal total dissolved solids (TDS) in feed solution—75,000 mg/L • Membrane scaling and fouling • Concentration polarization at the membrane • Chemicals used in the process discharged with the brine	• Membrane improvement • Development of new membrane materials • Role in hybrid systems	Dore (2005), Macedonio et al. (2012), World-Bank (2012), Goh et al. (2016), Tong and Elimelech (2016), Shahzad et al. (2017), and Vane (2017)
Electrodialysis (ED) and reverse electrodialysis (RED)	• Compared to RO: No applied pressure, higher water recovery, less membrane fouling and scaling due to RED, can operate at temperatures up to 50°C • Compared to thermal process: lower energy requirement and capital costs	• Cost-effective only for brackish waters • Removal of charged components only. Neutral components such as viruses and bacteria are not removed • Low electrode life due to corrosion	• Industrial waste water treatment	Strathmann (2010), Mezher et al. (2011), and Burn et al. (2015)

Continued

Table 2.1 Comparison among different desalination processes—cont'd

Process	Strengths	Limitations	Outlook	References
	• Less pretreatment due to RED • RED can produce energy			
Forward osmosis (FO) and pressure retarded osmosis (PRO)	• Driven by natural osmotic pressure gradient, without an externally applied pressure • Theoretically, requires less energy than RO and thermal processes • Can treat feedwater with high TDS (up to 175,000 mg/L) • Lower membrane fouling than RO. Fouling can be removed by osmotic backwashing, without the use of chemicals • Higher water recovery than RO and thermal processes • High rejection of trace organic contaminants and boron • PRO can produce energy • Can utilize waste heat for draw solution regeneration	• Uses a concentrated draw solution • Product water extraction and draw solution regeneration are energetically expensive • Control of osmotic pressure needed to prevent cross-contamination between feed and draw solution • Concentration polarization at the membrane • Lower flow rate than RO	• Emerging technology • Pilot and commercial plants • Optimization of draw solutions • Role in hybrid systems	Macedonio et al. (2012), Shaffer et al. (2015), Subramani and Jacangelo (2015), Goh et al. (2016), and Chua and Rahimi (2017)

Multistage flash distillation (MSF)	• Better than RO for high TDS feedwater • Commercially operating large-sized plants • Long-term operation record • Easy to manage and operate	• Higher energy requirement than RO and MED • Lower recovery than RO • Brine hotter by up to 15°C than receiving environment • Corrosion and scaling problems • Cannot operate under 60% capacity	• Not projected to expand	Fritzmann et al. (2007), Mezher et al. (2011), Macedonio et al. (2012), and Shahzad et al. (2017)
Multiple effect distillation (MED)	• Better than RO for high TDS feedwater • Commercially operating large-sized plants • Lower operation temperature than MSF • Can be incorporated with mechanical or thermal vapor compression system to increase efficiency • Can utilize waste heat or renewable energy sources	• Higher energy requirement than RO • More costly than MSF • Lower recovery than RO • Brine hotter by up to 15°C than receiving environment • Corrosion and scaling problems	• Projected to expand • Technological improvements and lower operating temperatures • Role in hybrid systems • Operation with renewable energy	Ophir and Lokiec (2005), Mezher et al. (2011), El-Ghonemy (2012), and Shahzad et al. (2017)

Continued

Table 2.1 Comparison among different desalination processes—cont'd

Process	Strengths	Limitations	Outlook	References
Membrane distillation (MD)	• High TDS feedwater (>200,000 mg/L) • Less membrane fouling than RO • Can utilize waste heat and renewable energy sources • High rejection rate and nonvolatile removal • Portable desalination plants	• Low water flux and recovery • Potential membrane wetting • Lack of specific MD membranes • Module more complex than RO • Uses electrical and thermal energy	• Emerging technology • Membrane engineering • Pilot and commercial plants • Role in hybrid systems	Subramani and Jacangelo (2015), Tong and Elimelech (2016), Amy et al. (2017), and Baghbanzadeh et al. (2017)

[a]See Further Reading for additional references.

manage the brine; and (c) treat the product water prior to its introduction into the water grid.

2.2.1 Pretreatment and Post-treatment

2.2.1.1 Pretreatment

Pretreatment in membrane processes is geared toward protecting the membranes from scaling and fouling by the diverse components present in the feedwater (see Chapter 3). Scaling, or inorganic fouling, is caused by the precipitation of inorganic salts, such as calcium carbonate, calcium sulfate, barium sulfate, and silicates, on the membrane. Salt precipitation occurs when their concentration in the brine exceeds their solubility. Organic fouling can develop by the accumulation of dissolved colloids and particulate organics, such as humic substances, on the surface or in the membrane pores. Biofouling, the most studied fouling effect, is defined as the adhesion of microorganisms on the membrane to form a biofilm that can further grow and entrap organic molecules, particles, and other microbial cells. A scaled or fouled membrane requires a higher-than-normal operating pressure to maintain the water flux, has a lower-than-expected salt rejection, and requires more frequent cleaning.

Pretreatment of the feedwater to reduce its scaling and fouling components can be physical, chemical and/or biological. Physical pretreatment includes, among others, filtration of feedwater through screens, sand filtration, multimedia pressure filtration, activated carbon filtration, zeolite filtration, microfiltration (MF, 0.1–10 μm), UF (0.01–0.1 μm), nanofiltration (NF, 1–10 nm), and dissolved air flotation.

Chemical pretreatment includes the addition of chemicals such as coagulants, flocculants, biocides, and antiscalants (scale inhibitors) to the feedwater at various stages along the desalination process. Coagulants are used to form larger particles from small charged particles present in the feedwater. They are typically inorganic salts of iron and aluminum or organic polymers such as polyamines. Flocculants, which are used to agglomerate noncharged particles, are molecules with high molecular-weight, such as anionic polyacrylamides and nonionic polymers. Both coagulation and flocculation are followed by a filtration step. Biocides, such as chlorine, ozone, and UV light, may be used to kill microorganisms and prevent biofouling. Any residual chlorine is oxidized, usually by sodium metabisulfite, to prevent damage to the membranes. Antiscalants prevent the deposition of inorganic salts on the membrane by allowing for supersaturation, by changing crystal shapes to produce nonadherent scales, or by imparting a highly negative charge to the crystals thereby keeping them separated, preventing propagation. Anti-scalants contain sulfonate, phosphate, phosphonate or carboxylic acid functional groups, polymaleic acid, polyacrylic acid, or a blend of them. Periodic chemical cleaning of the membranes is necessary and conducted using acidic and alkaline solutions, detergents, chelating agents as EDTA, and biocides.

Biological pretreatment uses microbial activity to break down biodegradable organics and nutrients in membrane bioreactors.

2.2.1.2 Post-treatment

The product water does not usually conform to drinking or agricultural water standards, due to the lack of essential minerals and its corrosiveness (low pH due to CO_2 presence) and has to be "re-hardened." Widely applied methods for remineralization and pH and alkalinity control are limestone ($CaCO_3$) or lime ($Ca(OH)_2$) dissolution aided by carbon dioxide. Blending with mineral-rich waters can also be used in several processes. In addition, biocides (usually chlorine) are used to protect consumers from possible contamination during storage and distribution of the product water.

The pretreatment and post-treatment methodologies previously described apply to most of the membrane processes discussed in Sections 2.2.2–2.2.5. Therefore, only process-specific methodologies will be addressed in the following sections.

2.2.2 Reverse Osmosis

RO is the most prevalent commercial desalination process used globally, comprising about 65% of the desalination industry. RO started to expand in the 1960s, with the development of commercially viable membranes by Loeb and Sourirajan (1963).

The process. RO uses pressure, powered by electrical or mechanical energy, at ambient temperature, to force water molecules from the feed solution through semipermeable membranes. The application of pressure is necessary to overcome the natural osmotic pressure gradient between the feed and the product water. The semipermeable membrane retains the salts and filter particles, separating water from salts, and produces product water and brine. The pressure used in the process ranges from 5.5 to 8.3 MPa for SW and is lower for BW, at 1.0–1.5 MPa. SWRO is usually run as one-pass system, operating at a single constant pressure determined by the brine's osmotic pressure, the water flux and the recovery. Some plants incorporate a second RO pass to reduce boron concentration to acceptable values for

Fig. 2.2 Reverse osmosis (RO) train consisting of an array of interconnected pressure vessels (A). The pressure vessels contain the RO membranes (B). Panel (C) shows a cross-section of a membrane and its spiral wound configuration. The Palmachim desalination plant in Israel, with a capacity to produce 246,575 m^3/day of fresh water uses about 25,000 membranes in the process.

drinking water and agricultural use. The strengths and limitations of RO are presented in Table 2.1.

Membranes. Commercial membranes for RO are usually nonporous thin film composites from polyamide and cellulose acetate. The membranes are configured as spiral-wound with spacers for feed and product water between the films (Fig. 2.2).

Efficiency and brine composition. The efficiency of the RO process varies based on the feedwater quality and on the setup of the desalination plant. On average, the recovery is 40%–50% for SWRO and 75% for BWRO. Consequently, brine from SWRO has about twice the salinity of SW and in BWRO, four times the salinity of BW. RO brine is discharged at ambient temperature and may include chemicals used along the process.

Energy requirements, cost and greenhouse gas (GHG) emissions. The energy required to desalinate SW at 50% recovery is currently 2–4 kWh/m^3 of product water (m^{-3} hereafter), down from 20 kWh/m^3 in the 1970s. BWRO requires less energy than SWRO (0.7–1 kWh/m^3). RO plants usually incorporate energy recovery devices using the pressurized brine exiting the membrane unit as the driving force. Production cost, following the energy reduction trend, has been reduced significantly since the mid-1990s and currently stands at \$0.3–1 m^{-3} for SWRO, depending on plant size and location. SWRO plants emit 1.4–2.8 kg CO_2 m^{-3} and 10–100 g NO_x m^{-3}.

Examples. RO plants constitute about 80% of the number of desalination plants globally, ranging from small- to mega-sized plants, with a typical production capacity of 128,000 m^3/day. One of the largest operating SWRO plant is Soreq (Israel, 411,000 m^3/day). Additional examples of SWRO are: Kwinana (Perth, Australia, 123,000 m^3/day), Carlsbad (California, United States, 123,000 m^3/day), Tuas (Singapore, 110,000 m^3/day), and Jedah Phases 1 and 2 (Kingdom of Saudi Arabia (KSA), 100,000 m^3/day). BW is desalinated by RO in El Paso (Texas, United States, 104,000 m^3/day).

2.2.3 Electrodialysis and Reverse Electrodialysis

ED is a process driven by electrical energy that separates salt ions from water. ED goes back >50 years but has been historically cost prohibitive. It constitutes only 3% of the global online desalination capacity and uses BW as feedwater. The prevalence of ED is expected to increase due to new, more cost-effective ion exchange membranes.

The process. In ED, two electrodes (a cathode and an anode), are separated by a stack of alternating anion and cation exchange membranes with solutions between them. When an electric potential gradient is applied across the electrodes, cations move toward the cathode, pass through the cation exchange membranes, and are retained by the anion exchange membranes. Similarly, anions move toward the anode, pass through the anion exchange membranes and are retained by the cation exchange membranes. This process yields two alternating streams: deionized product water and brine. ED removes only ionic components from the feedwater; therefore, membrane fouling by uncharged species like silica is less severe compared to RO. However, uncharged substances, such as viruses or bacteria, will go through to the product water. The strengths and limitations of ED are presented in Table 2.1.

Membranes. ED uses ion-exchange membranes that include a polymer matrix, ion groups fixed on the matrix, and mobile ions in the interstices. Cation-exchange membranes contain fixed negatively charged groups (such as sulfonic and carboxylic acids), and anion exchange membranes contain fixed positively charged groups (such as quaternary and tertiary amines).

Pretreatment. Typically, CO_2 is removed prior to ED to improve energy efficiency. Most ED desalination systems incorporate a step of RED, referred to as the "clean in place" technique. In RED, the polarity of the applied electrical potential is reversed, removing the charged particles that may have been precipitated on the membranes, thus reducing fouling and the use of chemicals.

Efficiency. A usual recovery of ED is 80% for BW.

Energy requirement and cost. The energy requirement for ED is proportional to the feedwater salinity: 3–7 kWh/m^3 for BW desalination and 17 kWh/m^3 for SW desalination. ED is cost-effective only for BW desalination ($0.6 m^{-3}).

Energy production. RED is an emerging process used to generate electrical energy from salinity gradients. When two solutions with different salinities flow through two cells separated by an ion exchange membrane, the salinity gradient generates a potential difference across it. When the sum of cell-pair voltages, which are additive, exceeds the electrode reaction potentials, an electrical current flows between the cathode and anode. The power produced can be utilized when connected to an external circuit.

Examples. Most ED plants are small- to medium-sized plants, for example in Sarasota (Florida, United States, 45,000 m^3/day). An exception is the large Abrera plant (Barcelona, Spain) that treats riverine water (200,000 m^3/day).

2.2.4 Forward Osmosis and Pressure Retarded Osmosis

Osmosis is a well-known natural physical phenomenon described in the 18th century by Nollet (1748) and of extreme importance in biology. In the desalination industry, FO and PRO are emerging technologies, not used yet as standalone processes to desalinate water, only as components in hybrid systems (see Section 2.4).

The process. FO is based on the principle that water diffuses through a semipermeable membrane from a low concentration solution, the feedwater, to a high concentration solution, the draw solution. The process is driven by the natural osmotic pressure gradient, at ambient temperature and without an external applied pressure. Following diffusion, the product water is separated from the draw solution and the latter regenerated. Draw solution regeneration can be an expensive process with costs depending on its characteristics. Most draw solutions are organic based but may also be inorganic based (such as NH_3-CO_2, NaCl, $MgCl_2$), micellar solutions, or magnetic nanoparticles. FO can be applied to high salinity feedwater and utilize waste heat to regenerate the draw solution. The strengths and limitations of FO are presented in Table 2.1.

Membranes. Membranes are specific for FO use and consist of flat-sheet cellulose triacetate, based on RO thin composite.

Efficiency and brine composition. The efficiency of FO for SW desalination is 35%, and brine generated is 1.5 times more concentrated than SW. In a hybrid FO-RO system, the efficiency can reach up to 95% (see Section 2.4).

Energy consumption. The energy requirements are similar for that of RO due to the energy needed for the draw solution regeneration step.

Energy production. PRO is an emerging process used more to generate energy than to produce fresh water. In PRO, a hydraulic counter-pressure is applied across the membrane at the draw solution side to retard but not prevent FO. Water permeates through the membrane, increasing the volume (and hence the pressure) of the draw solution in the pressurized chamber, which is used to generate power in a turbine. The most studied application for energy production by PRO is using riverine water and SW as feed and draw solutions, respectively.

2.2.5 Additional Membrane Processes

2.2.5.1 Nanofiltration

NF is a filtration processes that uses spiral-wound, polyamide-based membranes with a molecular weight cut-off between 200 and 1000 Da. NF membranes are useful to remove divalent ions from feedwater, such as calcium and magnesium. It also shows potential to retain pollutants and dissolved organic matter but like other membrane processes is prone to scaling and fouling. NF requires a lower operating pressure than RO but cannot be used for SW desalination, only for mildly BW. However, a hybrid system of NF-RO can be used to treat SW (see Section 2.4). The NF process is used to treat well water in Boca Raton (Florida, United States) at a capacity of 151,000 m^3/day.

2.2.5.2 Pervaporation

Pervaporation (PV) separates mixtures in contact with a membrane by preferential removal of a component due to its higher affinity or faster diffusion through the membrane. Several membrane materials have been evaluated in the laboratory: polyvinyl alcohol (PVA), hybrid organic-inorganic membranes based on PVA, maleic acid and silica, polyetheramide-based polymer. The main limitation of the process is its low water flux.

2.2.5.3 Membrane Capacitive Deionization

Capacitive deionization (CDI) is an electrosorption, two-step process, related to ED. A CDI cell consists of two electrodes separated by a spacer that acts as a flow channel for the feedwater. The electrodes are made out of activated carbon with high surface area. When an electrical potential is applied between the electrodes, ions are removed from the feed solution and adsorbed at the surface of the charged electrodes, producing deionized water. Upon saturation, the electrodes are regenerated by reversing the electric potential that releases the ions back into the solution, producing brine. In MCDI a cation exchange membrane is placed in front of the cathode and an anion exchange membrane is placed in front of the anode. The membranes prevent ion diffusion from the electrodes into the product water during the deionization step and ion adsorption into the electrodes during the regeneration step. MCDI has relatively low energy requirements: 0.1 kWh/m^3 for BW desalination and 1.8 kWh/m^3 for SW. This process is still at the bench and pilot stages.

2.3 THERMAL PROCESSES

Thermal processes are phase change processes in which feedwater is heated (under suitable operating temperatures and pressures) and the vapor condensed as pure water, leaving behind salts and other nonvolatile species. Historically, thermal distillation was formally described in the 17th century and used to desalinate SW aboard ships. Thermal processes are driven by thermal and mechanical energy and require more energy than membrane processes. Therefore, most processes operate with multiple steps, reusing heat through sequential condensation and evaporation. Established and emerging thermal processes include: multistage flash distillation (MSF), multieffect distillation (MED), membrane distillation (MD), adsorption desalination (AD), humidification dehumidification (HDH), freeze desalination (FD) (see Sections 2.3.2–2.3.5).

Overall, thermal processes have common operational goals: maximize production; minimize energy use; and prevent scaling, foaming, and corrosion. As for the membrane processes, thermal processes need to: (1) pretreat the feedwater, (2) manage the brine, and (3) treat the product water prior to its introduction into the water grid.

2.3.1 Pretreatment and Post-treatment

Pretreatment of the feedwater in thermal processes include, as for membrane processes, filtration and antiscalant application. In thermal processes, the feedwater is also pretreated to prevent foaming and corrosion. Chemicals used as antifoaming agents include polyethylene, polypropylene glycol, and detergents. Corrosion can be reduced by removing CO_2 from the feedwater by acid addition and by removing O_2 from the feedwater by sodium bisulfite addition. Corrosion and subsequent discharge of metals with the brine can be also reduced by coating corrosion-sensitive elements in the desalination plant with protective paints or resins. Post-treatments are similar to those described for membrane desalination (see Section 2.2.1.2).

The pretreatment and post-treatment methodologies previously described apply to most of the thermal processes discussed in Sections 2.3.2–2.3.5. Therefore, only process-specific methodologies will be addressed in the following sections.

2.3.2 Multistage Flash Distillation

MSF is an established thermal process used for desalination, and it encompasses 23% of the total global desalination. MSF technology dominated desalination up to the 1990s when it was surpassed by RO.

The process. In MSF, a series of stages (typically 18–25), each with successively lower temperature and pressure, are used to rapidly vaporize (or "flash") water from the feedwater. The vapor is then condensed by tubes of the inflowing cooler feedwater, recovering energy from the heat of condensation. MSF operates at top brine temperature ranging from 90°C to 110°C. Higher top brine temperature enhances flashing and performance ratio but increases scaling. The strengths and limitations of MSF are presented in Table 2.1.

Efficiency and brine composition. The process efficiency is on average 25% for SW but may range from 15% to 50%. The brine has about 1.2–1.5 times higher salinity than SW, and its temperature is 5–15°C higher than ambient. The brine may include chemicals used during the desalination process and corrosion products.

Energy requirement, cost, and GHG emissions. The typical energy consumption (thermal and electrical) for MSF ranges from 50 to 80 kWh/m^3 with cost in the range of \$0.56–1.76 m^{-3}. CO_2 emission is about 20 kg/m^3, and NOx emission is about 25 g/m^3.

Examples. MSF plants are common in, but not exclusive to, the Gulf area. For example, Shoaiba Phase 1 (KSA, 190,000 m^3/day), Ras Abu Fontas (Qatar, 160,000 m^3/day), Yanbu Phase 2 (KSA, 120,000 m^3/day), Jabel Ali L1 (Dubai, United Arab Emirates (UAE) 318,000 m^3/day), and Magtaa (Algeria, 500,000 m^3/day).

2.3.3 Multiple Effect Distillation

MED is an established thermal processes used in desalination, the first process to have been implemented commercially. MED encompasses 8% of the total global industry and is expected to increase due to recent technological improvements.

The process. MED is a thin-film evaporation process. The vapor produced from the feedwater in one chamber (or "effect") subsequently condenses in the next chamber, which exists at a lower temperature and pressure, providing additional heat of vaporization. MED operates at top brine temperatures of 65–70°C. Large MED plants incorporate thermal vapor compression (TVC) to improve efficiency by utilizing the pressure of the steam. The strengths and limitations of MSF are presented in Table 2.1.

Efficiency and brine composition. The process efficiency ranges from 15% to 50% for SW. The brine has salinity about 1.2–1.5 times higher than SW and temperature 5–15°C higher than ambient. The brine may include chemicals used during the desalination process and corrosion products.

Energy consumption, cost, and GHG emissions. The energy requirements for MED are lower than for MSF due to reduced pumping requirements. Production cost ranges from \$0.52 to \$1.02 m^{-3}. MED emits 7.0–17.6 kg/m^3 CO_2.

Examples. MED plants are common in, but not exclusive to, the Gulf area. For example: Az Zour (Kuwait, 486,000 m^3/day), Ras Laffan (Qatar, 286,000 m^3/day) Gujarat (India, 65,000 m^3/day), and Tianjin (Beijing, China, 200,000 m^3/day).

2.3.4 Membrane Distillation

MD is a thermally driven process that utilizes a hydrophobic, microporous membrane to separate water vapor from liquid. Although considered a hybrid technology, it is addressed here as a thermal due to the phase change of water. MD is at the bench or pilot scale.

The process. In MD, the driving force of the process is the partial vapor pressure difference maintained at the two interfaces of the membrane. Briefly, hot feedwater is brought into contact with the membrane, which allows only the vapor to pass through its pores. The vapor is then condensed on the coolant side. The process uses lower temperatures and pressures compared to the established thermal and membrane processes and can utilize low-grade waste heat or RE as energy sources. MD includes four configurations: (1) direct contact membrane distillation (DCMD) in which the membrane is in direct contact with liquid phases; (2) air gap membrane distillation (AGMD) in which an air gap is interposed between the membrane and a condensation surface; (3) vacuum membrane distillation (VMD) in which the permeate side is vapor or air under reduced pressure; and (4) sweep gas membrane distillation (SGMD) in which stripping gas is used as a carrier for the produced vapor. The strengths and limitations of MD are presented in Table 2.1.

Membrane. Common materials used for MD membranes are polytetrafluoroethylene, polyvinylidene-fluoride and polypropylene.

Efficiency and brine composition. Water recovery in MD can reach 90%. The brine, 10 times more saline than the feedwater, has a temperature 5–15°C higher than ambient.

Energy consumption and cost. MD consumes 43 kWh/m^3 or 10 kWh/m^3 if using waste heat. The estimated cost is \$1.2–2.0 m^{-3}.

Example. A small (10 m^3/day capacity) operates in the Maldives. Typical plant capacity is expected to be 24,000 m^3/day.

2.3.5 Additional Thermal Processes

2.3.5.1 Adsorption Desalination

AD is an adsorption-desorption cycle process driven by low heat, currently at the bench or pilot stage. In AD, feedwater enters an evaporator at ambient temperature and an adsorbent (highly porous silica gel) is used to adsorb the vapor generated at a low-pressure environment. When saturated, the adsorbent is heated to release the vapor (desorption process) that is then condensed inside an external condenser. AD can utilize low temperature waste heat or solar energy and has a low energetic requirement, <1.5 kWh/m^3, for SW desalination. The strengths and limitations of AD are given in Table 2.1.

2.3.5.2 Humidification Dehumidification

HDH utilizes the difference in water vapor pressure between a heated feed solution and an air stream to transport water across a membrane. The air stream is humidified with water permeating the membrane and dehumidified in a condenser using cold feedwater. HDH is a hybrid process still at the bench scale. It is simple to operate and maintain, has a high recovery ($>97\%$), high salt rejection, and lower operating temperatures compared to conventional thermal desalination. However, it is a high-energy process and is expected to be cost-effective only when utilizing waste heat or RE sources. The calculated energy consumption is 300–550 kWh/m^3 using conventional energy sources and 45 kWh/m^3 if utilizing waste heat or RE. HDH plants will require large areas and are expected to be suitable for small-scale remote applications when combined with RE.

2.3.5.3 Freeze Desalination

The FD process is based on water phase change, from liquid to solid. As ice crystals form, salt and other impurities are excluded from their structure. Residual salt is then washed from the crystals, recovering product water. The process is theoretically more efficient than thermal process, for the energy needed to freeze water is one seventh of that required to boil water. Although envisaged by Zarchin in the 1960s, it has not been applied commercially due to many technical difficulties. FD is now reemerging due to the prospect, among others, to use cold energy from regasification of liquefied natural gas (−162°C) as energy source.

2.4 HYBRID SYSTEMS

Hybrid systems usually include one desalination process and at least one additional process, the latter used to pretreat feedwater prior to desalination, to treat the brine prior to its management, and/or to produce energy. Table 2.2 compiles some of the possible hybrid systems at the operational, laboratory, and modeling stages with their relevant features.

2.5 APPROACHES TO IMPROVE DESALINATION

Many approaches are being pursued to improve desalination by increasing yield and selectivity, by reducing energy consumption and environmental footprint, by reducing wastes and the use of chemicals, and by increasing the utilization of waste heat and renewable energy. The following sections describe some of the most salient areas.

2.5.1 Membrane Engineering

Membrane engineering aims to improve existing membranes and develop new materials to enhance selectivity, permeability, and mechanical strength and to reduce fouling, scaling, and costs. In nanocomposite or nanoenhanced membranes, the existing polymer membrane matrix is incorporated with nanomaterials, such as metal/metal oxide (titanium dioxide, zeolites, silica, silver), carbon based nanotubes, and hydrophilic functional groups. Nanocomposite membranes have higher permeability than conventional RO membranes while maintaining salt rejection and require lower feed pressure but are more expensive than conventional RO membranes. They are currently applied for SWRO, with 40%–50% recovery and energy consumption of 1.7–2.5 kWh/m^3.

Table 2.2 Features and elements of potential hybrid systems; most systems are at the modeling or at the bench scale stage

Features	Potential hybrid systems	References
Reduction of brine volume, increased water recovery	RO-FO RO-MD	Camacho et al. (2013), Qasim et al. (2015), and Shaffer et al. (2015)
Reduction of brine volume, increased water recovery, energy generation	RO-RED RO-PRO	Kim et al. (2013) and Amy et al. (2017)
Dilution of feedwater, increased desalination efficiency	FO-RO PRO-RO NF-RO FO-ED-RO	NRC (2008), El-Ghonemy (2012), Qasim et al. (2015), Shaffer et al. (2015), Altaee et al. (2016), Bitaw et al. (2016), Goh et al. (2016), Amy et al. (2017), and Shahzad et al. (2017)
Dilution of feedwater, energy generation	FO-PRO	Altaee et al. (2016), Goh et al. (2016), and Amy et al. (2017)
Draw solution regeneration	FO-MD	Camacho et al. (2013), Shaffer et al. (2015), Goh et al. (2016), and Amy et al. (2017)
Product water treatment and brine concentration	MD-RED MD-PRO	Amy et al. (2017)
Latent heat of thermal process used for MD to increase water recovery	MSF-MD MED-MD	Amy et al. (2017)
Optimization of energy use and water production	MSF-RO MED-RO Operated in parallel, using energy from a co-located power plant	Commercial plants, for example. Fujairah 1 and Fujairah 2 plants in the United Arab Emirates

ED, electrodialysis; *FO*, forward osmosis; *MD*, membrane distillation; *MED*, multieffect distillation; *MSF*, multistage flash distillation; *PRO*, pressure retarded osmosis; *RED*, reverse electrodialysis; *RO*, reverse osmosis.

Emerging new materials are the nanostructured membranes in which nano-sized pores constitute their internal structure. Among them are the Fullerene (C_{60})-based nanomaterials such as carbon nanotubes, graphene-based nanomaterials, and biomimetic aquaporin (AQP). AQP permeability is one order of magnitude higher than commercial RO membranes, and nanotubes membranes permeability is 10 times higher. The extensive research of nanostructured membranes is only at bench scale level with limited data on real feedwater. In addition, health risks associated with release of nanomaterials into treated water are unknown.

2.5.2 Zero liquid discharge

Zero liquid discharge (ZLD) is a management concept and a technological challenge recently expanding in the desalination industry. ZLD aims to eliminate brine discharge while recovering water and salts and protecting the environment. Early ZLD, still used today, is based on thermal processes where the brine or waste water are evaporated in a brine concentrator followed by a brine crystallizer, an evaporation pond, or a wind-aided intensified evaporation tower. Water is recovered, and the residual salt is either reused or disposed of at landfills. The process is important for inland desalination, but it is costly having high energy requirements and GHG emissions. Currently, RO, ED, and the emerging FO and MD are being considered as replacements for the thermal processes to achieve ZLD. Still, capital and operating costs are seen as prohibitive, and regulatory incentives are necessary for ZLD implementation.

2.5.3 Novel Desalination Technologies

2.5.3.1 Microbial Desalination Cell

Microbial desalination cell (MDC) is similar to the ED process but uses the energy produced by microbial oxidation of organic matter in a microbial fuel cell (MFC) instead of electrical energy as in ED. The MDC system consists of three chambers: one with an anode, one with a cathode, and a middle chamber with the feedwater. The middle chamber is separated from the anode by an anion exchange membrane and from the cathode by a cation exchange membrane. Bacteria oxidize organic matter at the anode and oxygen is reduced at the cathode, inducing the movement of anions toward the anode and cations toward the cathode, leaving desalinated water in the middle chamber. The efficiency of the process is low and MCD is envisaged as an RO pretreatment and not a standalone process.

2.5.3.2 Ion Concentration Polarization

Ion concentration polarization (ICP) is an electro-kinetic phenomenon created by selective charge transport, usually when ions pass through an ion-selective membrane. ICP is known to hinder the RO process (see Section 2.2), but it is now being pursued as an emerging technology for water treatment and desalination. An ICP cell is composed of two parallel microfluidic compartments linked by a perpendicular nanofluidic channel. When a voltage is applied across the nanochannel, ions are transported creating an ion depletion zone (IDZ). For desalination, charged species are continuously diverted into a side channel (the brine stream) for removal. ICP can also exclude large particles, such as viruses and microorganisms, from the product water. However, the efficiency of the process is very low. MDC is still at the laboratory scale. It is envisaged for small-scale systems in remote locations, possible battery-powered.

2.5.3.3 Clathrate Hydrates

Desalination by Clathrate hydrates (CHs) formation is similar to FD where CHs, not ice, are formed. CHs are a crystalline inclusion of water and a guest molecule, such as methane, propane, hydrochlorofluorocarbon refrigerants. CHs have cage-like substructures that form spontaneously at a temperature (typically slightly above the freezing point of water) and pressure specific to each guest molecule. After occupation of a sufficient number of cages, a thermodynamically stable crystalline unit cell structure (gas hydrate) is formed. As in FD, the hydrates are separated from the brine and washed to remove the salt entrapped in interstitial spaces. This technology shows promise, but it is still at the research stage.

2.5.4 Renewable Energies for Desalination

The use of renewable energy to produce electricity and heat has been increasing steadily over the past decade and is addressed also by the desalination industry. The reasons to use RE in desalination are the high energy requirements for the processes (30%–75% of the operational costs), the need to reduce GHG emissions, and the need to increase sustainability. The proposed RE sources for desalination are similar to those for power generation: solar (thermal and photovoltaic), geothermal, wind, and wave (tidal) energies. In addition, SW temperature differences (ocean thermal energy) may be used to power desalination in the tropics, where large differences in temperature between surface and deeper water are found. An example for the latter is a small plant operating in Kavaratti, India. Warm surface waters

(25–30°C) are pumped into a vacuum chamber where the water vaporizes. The vapor is separated by MD and condensed in a separate chamber, cooled by colder deep waters (13°C). As of 2015, only 1% of the total global desalination plants were using RE sources: 43% solar photovoltaic, 27% solar thermal, 20% wind, and 10% hybrid. The International Atomic Energy Agency forecasts that by 2030 RE-powered desalination will be sufficient only for domestic water supply but will expand to meet industrial supply by 2050. The use of nuclear power for desalination is being assessed as well.

REFERENCES

Altaee, A., Millar, G.J., Zaragoza, G., 2016. Integration and optimization of pressure retarded osmosis with reverse osmosis for power generation and high efficiency desalination. Energy 103, 110–118.

Amy, G., Ghaffour, N., Li, Z., Francis, L., Linares, R.V., Missimer, T., Lattemann, S., 2017. Membrane-based seawater desalination: present and future prospects. Desalination 401, 16–21.

Baghbanzadeh, M., Rana, D., Lan, C.Q., Matsuura, T., 2017. Zero thermal input membrane distillation, a zero-waste and sustainable solution for freshwater shortage. Appl. Energy 187, 910–928.

Bitaw, T.N., Park, K., Yang, D.R., 2016. Optimization on a new hybrid forward osmosis-electrodialysis-reverse osmosis seawater desalination process. Desalination 398, 265–281.

Burn, S., Hoang, M., Zarzo, D., Olewniak, F., Campos, E., Bolto, B., Barron, O., 2015. Desalination techniques—a review of the opportunities for desalination in agriculture. Desalination 364, 2–16.

Camacho, L., Dumée, L., Zhang, J., Li, J.-d., Duke, M., Gomez, J., Gray, S., 2013. Advances in membrane distillation for water desalination and purification applications. Water 5, 94.

Chua, H.T., Rahimi, B., 2017. Low Grade Heat Driven Multi-Effect Distillation and Desalination. Elsevier Science, Amsterdam, Netherlands.

Dore, M.H.I., 2005. Forecasting the economic costs of desalination technology. Desalination 172, 207–214.

El-Ghonemy, A.M.K., 2012. Future sustainable water desalination technologies for the Saudi Arabia: a review. Renew. Sust. Energ. Rev. 16, 6566–6597.

Fritzmann, C., Löwenberg, J., Wintgens, T., Melin, T., 2007. State-of-the-art of reverse osmosis desalination. Desalination 216, 1–76.

Goh, P.S., Matsuura, T., Ismail, A.F., Hilal, N., 2016. Recent trends in membranes and membrane processes for desalination. Desalination 391, 43–60.

Kim, J., Park, M., Snyder, S.A., Kim, J.H., 2013. Reverse osmosis (RO) and pressure retarded osmosis (PRO) hybrid processes: Model-based scenario study. Desalination 322, 121–130.

Loeb, S., Sourirajan, S., 1963. Sea water demineralization by means of an osmotic membrane. In: Saline Water Conversion—II. American Chemical Society, pp. 117–132.

Macedonio, F., Drioli, E., Gusev, A.A., Bardow, A., Semiat, R., Kurihara, M., 2012. Efficient technologies for worldwide clean water supply. Chem. Eng. Process. Process Intensif. 51, 2–17.

Mezher, T., Fath, H., Abbas, Z., Khaled, A., 2011. Techno-economic assessment and environmental impacts of desalination technologies. Desalination 266, 263–273.

Nollet, J.A., 1748. Leçons de physique expérimentale Hippolyte-Louis Guerin & Louis-Francios Delatour.
NRC, 2008. Desalination, a National Perspective National Research Council of the National Academies. The National Academies press, Washington, DC.
Ophir, A., Lokiec, F., 2005. Advanced MED process for most economical sea water desalination. Desalination 182, 187–198.
Qasim, M., Darwish, N.A., Sarp, S., Hilal, N., 2015. Water desalination by forward (direct) osmosis phenomenon: a comprehensive review. Desalination 374, 47–69.
Shaffer, D.L., Werber, J.R., Jaramillo, H., Lin, S., Elimelech, M., 2015. Forward osmosis: where are we now? Desalination 356, 271–284.
Shahzad, M.W., Burhan, M., Ang, L., Ng, K.C., 2017. Energy-water-environment nexus underpinning future desalination sustainability. Desalination 413, 52–64.
Strathmann, H., 2010. Electrodialysis, a mature technology with a multitude of new applications. Desalination 264, 268–288.
Subramani, A., Jacangelo, J.G., 2015. Emerging desalination technologies for water treatment: a critical review. Water Res. 75, 164–187.
Tong, T., Elimelech, M., 2016. The global rise of zero liquid discharge for wastewater management: drivers, technologies, and future directions. Environ. Sci. Technol. 50, 6846–6855.
Vane, L.M., 2017. Water recovery from brines and salt-saturated solutions: operability and thermodynamic efficiency considerations for desalination technologies. J. Chem. Technol. Biotechnol. 92, 2506–2518.
World-Bank, 2012. Renewable Energy Desalination: An Emerging Solution to Close the Water Gap in the Middle East and North Africa. World-Bank, Washington, DC.

FURTHER READING

Ang, W.L., Mohammad, A.W., Hilal, N., Leo, C.P., 2015. A review on the applicability of integrated/hybrid membrane processes in water treatment and desalination plants. Desalination 363, 2–18.
Anon, 1685. The Certificates of Several Captains and Masters of Ships, and Others, Both at Sea and Land, Who Have Used the Patentees Engine for Making Salt Water Fresh. Printed by John Harefinch in Mountague-Court in Little Britain, London.
Bhattacharjee, Y., 2007. Turning Ocean water into rain. Science 316, 1837.
Buonomenna, M.G., 2013. Nano-enhanced reverse osmosis membranes. Desalination 314, 73–88.
Burn, S., Gray, S., 2016. Efficient Desalination by Reverse Osmosis. A guide to RO practice. IWA, London.
Chang, J., Zuo, J., Lu, K.-J., Chung, T.-S., 2016. Freeze desalination of seawater using LNG cold energy. Water Res. 102, 282–293.
Edzwald, J.K., Haarhoff, J., 2011. Seawater pretreatment for reverse osmosis: Chemistry, contaminants, and coagulation. Water Res. 45, 5428–5440.
Elimelech, M., Phillip, W.A., 2011. The future of seawater desalination: energy, technology, and the environment. Science 333, 712–717.
Fitzgerald, R., 1683. Salt-Water Sweetned, or, A True Account of the Great Advantages of This New Invention Both by Sea and by Land Together With a Full and Satisfactory Answer to all Apparent Difficulties: Also the Approbation of the Colledge of Physicians: Likewise a Letter of the Honourable Robert Boyle to a Friend Upon the Same Subject. Printed for Will. Cademan, London.

Ghaffour, N., Missimer, T.M., Amy, G.L., 2013. Technical review and evaluation of the economics of water desalination: current and future challenges for better water supply sustainability. Desalination 309, 197–207.
Giwa, A., Dufour, V., Al Marzooqi, F., Al Kaabi, M., Hasan, S.W., 2017. Brine management methods: recent innovations and current status. Desalination 407, 1–23.
Goh, P.S., Matsuura, T., Ismail, A.F., Ng, B.C., 2017. The water–energy Nexus: solutions towards energy-efficient desalination. Energy Technol. 5, 1136–1155.
Greenlee, L.F., Lawler, D.F., Freeman, B.D., Marrot, B., Moulin, P., 2009. Reverse osmosis desalination: water sources, technology, and today's challenges. Water Res. 43, 2317–2348.
Gude, V.G., 2016. Desalination and sustainability—an appraisal and current perspective. Water Res. 89, 87–106.
Helfer, F., Lemckert, C., Anissimov, Y.G., 2014. Osmotic power with pressure retarded osmosis: theory, performance and trends—a review. J. Membr. Sci. 453, 337–358.
IAEA, 2015. New technologies for seawater desalination using nuclear energy. IAEA-TECDOC Series No 1753, International Atomic Energy Agency.
Imbrogno, J., Keating Iv, J.J., Kilduff, J., Belfort, G., 2017. Critical aspects of RO desalination: a combination strategy. Desalination 401, 68–87.
Khalifa, A., Ahmad, H., Antar, M., Laoui, T., Khayet, M., 2017. Experimental and theoretical investigations on water desalination using direct contact membrane distillation. Desalination 404, 22–34.
Khan, M.T., Manes, C.-L.D.O., Aubry, C., Croué, J.-P., 2013. Source water quality shaping different fouling scenarios in a full-scale desalination plant at the Red Sea. Water Res. 47, 558–568.
Khan, S.U.-D., Khan, S.U.-D., Haider, S., El-Leathy, A., Rana, U.A., Danish, S.N., Ullah, R., 2017. Development and techno-economic analysis of small modular nuclear reactor and desalination system across Middle East and North Africa region. Desalination 406, 51–59.
Kim, S.J., Ko, S.H., Kang, K.H., Han, J., 2010. Direct seawater desalination by ion concentration polarization. Nat. Nanotechnol. 5, 297–301.
Kim, Y.-D., Thu, K., Ng, K.C., Amy, G.L., Ghaffour, N., 2016. A novel integrated thermal-/membrane-based solar energy-driven hybrid desalination system: concept description and simulation results. Water Res. 100, 7–19.
Kress, N., Galil, B., 2016. Impact of seawater desalination by reverse osmosis on the marine environment. In: Burn, S., Gray, S. (Eds.), Efficient Desalination by Reverse Osmosis. IWA, London, pp. 177–202.
Kucera, J., 2015. Reverse Osmosis. Industrial Processes and Applications, second ed. John Wiley & Sons, Inc.
Lattemann, S., 2010. Development of an Environmental Impact Assessment and Decision Support System for Seawater Desalination Plants. (Ph.D. dissertation). Delft University of Technology, The Netherlands, p. 294.
Lattemann, S., Hopner, T., 2008. Environmental impact and impact assessment of seawater desalination. Desalination 220, 1–15.
Li, M., Anand, R.K., 2016. Recent advancements in ion concentration polarization. Analyst 141, 3496–3510.
Lin, S., Elimelech, M., 2017. Kinetics and energetics trade-off in reverse osmosis desalination with different configurations. Desalination 401, 42–52.
Lior, N., 2017. Sustainability as the quantitative norm for water desalination impacts. Desalination 401, 99–111.
Loeb, S., 1998. Energy production at the Dead Sea by pressure-retarded osmosis: challenge or chimera? Desalination 120, 247–262.

Logan, B.E., Elimelech, M., 2012. Membrane-based processes for sustainable power generation using water. Nature 488, 313–319.
Manes, C.L.D., West, N., Rapenne, S., Lebaron, P., 2011. Dynamic bacterial communities on reverse-osmosis membranes in a full-scale desalination plant. Biofouling 27, 47–58.
Meerganz von Medeazza, G.L., 2005. "Direct" and socially-induced environmental impacts of desalination. Desalination 185, 57–70.
Mehanna, M., Saito, T., Yan, J., Hickner, M., Cao, X., Huang, X., Logan, B.E., 2010. Using microbial desalination cells to reduce water salinity prior to reverse osmosis. Energy Environ. Sci. 3, 1114–1120.
Nebbia, G., Menozzi, G.N., 1968. Early experiments on water desalination by freezing. Desalination 5, 49–54.
Ng, K.C., Thu, K., Kim, Y., Chakraborty, A., Amy, G., 2013. Adsorption desalination: an emerging low-cost thermal desalination method. Desalination 308, 161–179.
Park, K.-N., Hong, S.Y., Lee, J.W., Kang, K.C., Lee, Y.C., Ha, M.-G., Lee, J.D., 2011. A new apparatus for seawater desalination by gas hydrate process and removal characteristics of dissolved minerals (Na^+, Mg^{2+}, Ca^{2+}, K^+, B^{3+}). Desalination 274, 91–96.
Perez-Gonzalez, A., Urtiaga, A.M., Ibanez, R., Ortiz, I., 2012. State of the art and review on the treatment technologies of water reverse osmosis concentrates. Water Res. 46, 267–283.
Post, J.W., Veerman, J., Hamelers, H.V.M., Euverink, G.J.W., Metz, S.J., Nymeijer, K., Buisman, C.J.N., 2007. Salinity-gradient power: evaluation of pressure-retarded osmosis and reverse electrodialysis. J. Membr. Sci. 288, 218–230.
Straub, A.P., Deshmukh, A., Elimelech, M., 2016. Pressure-retarded osmosis for power generation from salinity gradients: is it viable? Energy Environ. Sci. 9, 31–48.
Sutzkover-Gutman, I., Hasson, D., 2010. Feedwater pretreatment for desalination plants. Desalination 264, 289–296.
Turek, M., Bandura, B., 2007. Renewable energy by reverse electrodialysis. Desalination 205, 67–74.
Wang, P., Chung, T.-S., 2012. A conceptual demonstration of freeze desalination–membrane distillation (FD–MD) hybrid desalination process utilizing liquefied natural gas (LNG) cold energy. Water Res. 46, 4037–4052.
Wang, D.-X., Su, M., Yu, Z.-Y., Wang, X.-L., Ando, M., Shintani, T., 2005. Separation performance of a nanofiltration membrane influenced by species and concentration of ions. Desalination 175, 219–225.
WHO, 2007. Desalination for safe water supply. Guidance for the health and environmental aspects applicable to desalination. World Health Organization WHO/SDE/WSH/07.
Yaroshchuk, A., 2017. "Breakthrough" osmosis and unusually high power densities in pressure-retarded osmosis in non-ideally semi-permeable supported membranes. Sci. Rep. 745168.
Zarchin, A., 1963. Processes for Sweetening Saltwater by Freezing. U.S. Patent No. 3,093,975. 18 June 1963.

CHAPTER 3

Seawater Quality for Desalination Plants

The quality of seawater, the feedwater for seawater desalination, and the organisms present in the marine environment are of utmost importance for the efficient and reliable operation of a desalination plant. Seawater withdrawn from areas with poor or inconsistent quality will require extensive pretreatment prior to desalination. Desalination plants with intakes located near possible pollution sources will be prone to occasional plant closures, increasing costs and reducing the reliability of product water supply. Intakes placed in habitats with many organisms that can be entrained, impinged, and entrapped (see Chapter 4) may hinder desalination as well as impact the ecosystem.

This chapter is written with two objectives: (1) to describe concisely the marine environment (seawater composition, marine habitats, and marine organisms); and (2) to present the effects of the marine environment on plant operations, illustrated with actual examples when freshwater production was reduced or temporarily stopped. This chapter also sets the basis for the discussions on the potential and actual effects of desalination on the marine environment (Chapters 4–6).

3.1 SEAWATER COMPOSITION (EXCLUDING LIVING ORGANISMS)

The constituents of seawater can be classified by their concentrations: major (with concentrations higher than 100 mg/kg), minor (with concentrations from 1 to100 mg/kg) and trace (with concentrations lower than 1 mg/kg); by their chemical character: inorganic or organic; and by their physical state: dissolved (<0.2 μm diameter), colloidal or particulate (>0.2 μm diameter). In here, the physical state is the main characteristic used to divide among the constituents. A dissolved substance is truly in solution, homogeneously dispersed, and cannot be removed by physical methods such as filtration, sedimentation, and centrifugation. Particles are large enough to settle out of solution or be removed by filtration. Colloids are in the size range of 0.01–1 μm diameter,

Marine Impacts of Seawater Desalination: Science, Management, and Policy
https://doi.org/10.1016/B978-0-12-811953-2.00003-7

overlapping dissolved and particulate states. Like dissolved substances, colloids do not settle out of solution but can aggregate to form particles. Seawater constituents originate from natural and anthropogenic sources, both potentially influencing the desalination process.

3.1.1 Dissolved Constituents

3.1.1.1 Major Inorganic Constituents

The major constituents in seawater, sometimes called the conservative elements, are inorganic ions. The two most abundant ions are chloride (Cl^-) and sodium (Na^+), constituting 86% of all sea salts. Together with five additional major ions—sulfate (SO_4^{-2}), magnesium (Mg^{+2}), calcium (Ca^{+2}), potassium (K^+), and bicarbonate (HCO_3^-)—they make up about 99% of all sea salts (Table 3.1). These major constituents determine the salinity of seawater and are those that need to be removed to produce fresh water. Salinity is defined as the amount of salts (in grams) dissolved in 1 kg of seawater. Several units are in use for salinity in the literature: per mille (‰), part

Table 3.1 Dissolved inorganic constituents and gases in seawater with average salinity (S) of 35

A. Major and minor inorganic constituents and gases

Component	Units	Concentration at S = 35	Relative concentration (%)
Major			
Cl^-	g/kg	19.4	55.07
Na^+	g/kg	10.8	30.62
SO_4^{-2}	g/kg	2.71	7.72
Mg^{+2}	g/kg	1.29	3.68
Ca^{+2}	g/kg	0.41	1.17
K^+	g/kg	0.4	1.1
HCO_3^-	g/kg	0.11	0.3
Minor			
Br^-	g/kg	0.067	0.19
Sr^{+2}	g/kg	0.008	0.02
B^{+3}	g/kg	0.0045	0.013
F^-	g/kg	0.001	0.004

Component	Units	Concentration at S = 35	Units	Concentration at S = 35
Gases, in equilibrium with the atmosphere and seawater temperature of 20°C				
N_2	mg/L	11.7	µmol/kg	420
O_2	mg/L	7.23	µmol/kg	226

Table 3.1 Dissolved inorganic constituents and gases in seawater with average salinity (S) of 35—cont'd

Component	Units	Concentration at S = 35	Units	Concentration at S = 35
Ar	mg/L	0.44	μmol/kg	11.1
CO_2	mg/L	<0.51	μmol/kg	<11.6
Kr	mg/L	2×10^{-4}		
Ne	mg/L	1.2×10^{-4}		
Xe	mg/L	5×10^{-5}		
He	mg/L	6.8×10^{-6}		

B. *Trace constituents*. Nearly all the elements of the periodic table are found in the sea, most in trace concentrations. Below are the concentrations of the essential nutrients (*) and five additional elements

Component	Units	Concentration average
*Nutrients (N, P, and Si)	Nutrient concentrations vary spatially, temporally, and with water depth. Commonly, nutrient concentrations decrease with increased distance from the shoreline and increase with increasing water depth. Surficial waters in the open oceans are usually depleted from nutrients due to biological consumption during photosynthesis. Higher concentrations are found near pollution sources	
*N-(mainly NO_3^-)	mg/L	0.5
*P (PO_4^{-3})	mg/L	$6 \times\times 10^{-2}$
*Si-($Si(OH)_4$)	mg/L	2
Cu	mg/L	1×10^{-4}
Fe	mg/L	5.5×10^{-5}
Pb	mg/L	2×10^{-6}
Hg	mg/L	1×10^{-6}
Au	mg/L	2×10^{-8}

per thousand (ppt, g/kg), practical salinity unit (PSU), practical salinity scale (PSS). Salinity is also expressed without units, as it is usually measured as the ratio between the conductivity of a seawater sample and that of standard KCl solution. However, all salinity units are essentially numerically equal for the purpose of this book. The average salinity in the open ocean is 35. Salinity decreases with the addition of water and through precipitation and riverine runoff, and it increases with water removal by evaporation. For example, salinities in semienclosed seas located in arid regions exceed average ocean salinity: the Red Sea and the Gulf have 40–42 salinity, and in the Mediterranean Sea the salinity is about 38–39. Those are also the areas where the main global seawater desalination effort is concentrated.

However, while salinity varies, the relative concentrations (the ratios) of the major ions in seawater remain constant (Table 3.1). This is called the principle of constant proportion, or Dittmar's Principle, which was determined empirically by the analysis of the major elements in a large number of seawater samples. Small variations in the relative concentrations of calcium and bicarbonate may occur due to biological processes. Therefore, if the salinity of a seawater sample is measured, the concentrations of the major constituents can be calculated using the constant proportions. The density of a surficial seawater sample can be calculated by an empirical formula using its salinity and temperature. For seawater located further down the water column, pressure is an additional parameter added to the calculation.

The energy expenditure of a desalination process increases with increase in feedwater salinity. Therefore, if seawater salinity increases as a result of climate change or a result of brine discharge, it will increase production costs. The Gulf States are on the way to reach a "peak salt" when water may became so saline that desalination will not make economic sense and it will be too expensive to desalinate.

3.1.1.2 Minor Inorganic Constituents

The minor dissolved inorganic constituents in seawater include the ions of bromine, boron, strontium, and fluorine (Br^-, B^{+3}, Sr^{+2}, and F^-, respectively; see Table 3.1). The principle of constant proportion applies also to the minor inorganic constituents. Seawater desalination plants using membrane technology usually have to add an additional treatment to reduce boron concentrations to comply with drinking water quality criteria, usually <2.4 mg/L (see Chapter 1).

3.1.1.3 Trace Inorganic Constituents

Nearly all the elements of the periodic table are found in the sea, such as mercury, lead, and gold, most in trace concentrations (<1 mg/kg). Among the trace constituents of seawater are the inorganic nutrients, essential for plant and animal growth: nitrogen species (nitrate, nitrite, and ammonium), phosphate, iron, and copper, and for some organisms also silica (Table 3.1). They are supplied naturally to the oceans by land runoff, ground water intrusion and atmospheric deposition but also from anthropogenic sources. Nutrient concentrations vary spatially, temporally, and with water depth. Commonly, nutrient concentrations decrease with increased distance from the shoreline and increase with increasing water depth. Nutrients in seawater

A. *Photosynthesis, respiration and decomposition.* Photosynthesis occurs only in the euphotic zone, while respiration and decomposition occurs everywhere. Excess nutrients added to the photic zone may promote harmful algal blooms (HAB) due to increase in photosynthesis followed by depletion of oxygen due to bacterial decay of the dead biomass.

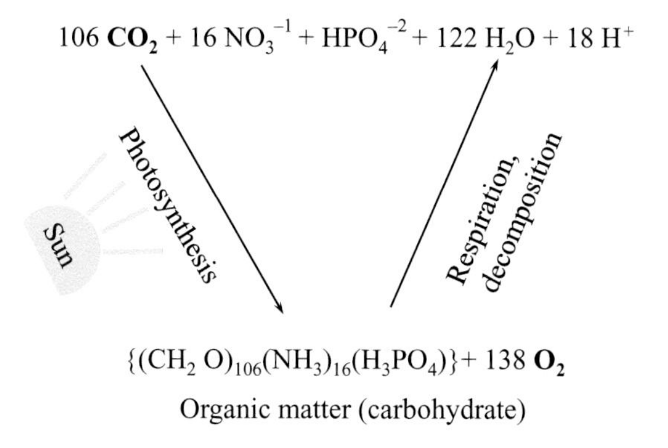

B. *Nitrogen fixation (mediated by autotrophic bacteria).* N_2 is fixed to alleviate limitation of productivity by the lack of N.

$N_2(aq) \longrightarrow NH_4^+$

C. *Nitrification (mediated by ammonia-oxidizing bacteria, in aerobic environments).*

$NH_4^+ \longrightarrow NO_2^- \longrightarrow NO_3^-$

D. Denitrification *(mediated by bacteria, in anoxic environments).* Nitrate replaces dissolved oxygen as an oxidizer.

$NO_3^- \longrightarrow N_2O \longrightarrow N_2(g)$

Fig. 3.1 Biological processes involving essential nutrients and gases dissolved in seawater.

are nonconservative, and their concentrations change following biological processes: decrease with photosynthesis and increase with respiration and decomposition of organic matter (Fig. 3.1). Excess nutrient supplied to the marine environment can promote anomalous and harmful algal blooms (HABs), consequently impairing desalination operations (see Section 3.3.1.1).

3.1.1.4 Dissolved Organic Matter

Dissolved organic matter (DOM) concentrations in seawater range from ng/L to mg/L. Natural DOM is comprised of compounds produced by organisms in the marine environment: metabolic by-products secreted by plants such as organic acids and sugars; of organic compounds excreted by animals, such as organic phosphorus and amino acids; and of organic compounds originating from the decomposition of dead organisms by bacteria, such as low molecular weight polymers. Natural oil seeps may also supply dissolved organic compounds to seawater. DOM can also have a natural terrestrial origin and be introduced to the sea by runoff, such as humic and fulvic acids, organometallic compounds, amino acids, and lipids. Anthropogenic sources can add a large array of DOM to seawater, some of them synthetic compounds that do not exist in nature. A known example is the pesticide DDT, which was banned from use or restricted. Recently, there are many emerging pollutants, such as pharmaceuticals, antiseptics, and personal care compounds that are raising concern about their effect on the environment. Within the context of desalination, high DOM may increase fouling propensity while specific pollutants, such as oil and BTEX, may damage the membranes.

3.1.1.5 Dissolved Gases

The major dissolved gases in seawater are nitrogen, oxygen, argon, and to a lesser degree, carbon dioxide (N_2, O_2, Ar, and CO_2, respectively) (Table 3.1). Other gases include noble gases (helium, neon, krypton, xenon), methane (CH_4), hydrogen sulfide (H_2S), radon, nitrous oxide, and dimethylsulfide. The sources of the gases to the ocean may be natural, through equilibrium with the atmosphere at the air-sea interface, from hydrothermal vents and gas seeps, or from in situ biological processes in the water column or in the sediments (Fig. 3.1). Gases may be also of anthropogenic origin such as chlorofluorocarbons (CFCs; i.e., Freon) and sulfur hexafluoride (SF_6). CO_2 and CH_4 have both anthropogenic and natural sources.

The solubility of gases in seawater increases with decreasing temperature and salinity, that is, gases dissolve more readily in cold, low salinity water. Each gas has a unique saturation value that can be calculated from an empirical formula. For example, the saturation of O_2 in seawater with 35 salinity at 20°C is 226 μmol/kg (7.43 mg/L) and higher at 10°C, 275 μmol/kg (9.07 mg/L). When the concentration of dissolved oxygen is lower than 2 mg/L, the seawater is termed hypoxic, and when no dissolved oxygen is present, the seawater is anoxic.

In addition to the effect of temperature and salinity, the concentration of dissolved gases in seawater may be changed by hydrographic processes of mixing, diffusion, advection, and, for some gases, biological activity.

Photosynthesis, respiration and organic matter decomposition change the concentrations of CO_2 and O_2 in seawater. N_2 concentrations may be changed by nitrogen fixation and nitrification/denitrification processes (Fig. 3.1).

CO_2 is unique among the gases for its dissolution in seawater is controlled also by chemical reactions (Eq. 3.1). When CO_2 dissolves in seawater, it hydrates to yield carbonic acid (H_2CO_3), which further dissociates to bicarbonate (HCO_3^-) and carbonate (CO_3^{-2}), forming a stable buffering system, the carbonate system. The main species of the carbonate system present in seawater is bicarbonate (90%). Dissolved CO_2 constitutes <1% of the inorganic carbon in seawater and carbonate <10%.

$$CO_2 + H_2O \leftrightarrow H_2CO_3 \leftrightarrow HCO_3^- + H^+ \leftrightarrow CO_3^{2-} + 2H^+ \quad (3.1)$$

The carbonate system keeps oceanic pH relatively constant at 7.5–8.5. The atmospheric concentration of CO_2 has been increasing since the industrial revolution due to fossil fuel burning and deforestation, from 280 ppm to >400 ppm currently (Hawaii's Mauna Loa Observatory, NOAA). The increase in atmospheric CO_2 has concurrently increased its concentration in the oceans leading to ocean acidification, reducing seawater pH and affecting the carbonate system.

3.1.2 Colloids

Colloids or submicrometer compounds (1 nm–1 μm diameter) are at the boundary between dissolved and particulate matter, overlapping with both. Similar to the dissolved components, colloids do not sink and are transported with the water parcel. However, they can agglomerate and coagulate and form aggregates. Colloids in seawater may be inorganic (such as iron oxyhydroxides) but are mostly derived from DOM, such as humic acid. Recently, the role of the colloidal transparent exopolymer particles (TEP) has been widely recognized in the formation of large aggregates. TEP are composed of acidic polysaccharides formed from DOM released by phytoplankton. Within the context of seawater desalination, TEP serve as the basis for biofilm formation, providing surfaces for the initial colonization by bacteria followed by biofouling by organisms at higher trophic levels (see Chapter 2).

3.1.3 Particulate Matter

Particulate matter in seawater is comprised of particles with a diameter larger than 0.2 μm, although 0.45 μm is also used as the cut-off size to define particles. Particulate matter may be of abiotic or biotic origin and originate from natural and anthropogenic sources. Abiotic particulate matter consists of minerals, such as clay and insoluble hydrous metal compounds, supplied to the marine environment naturally as atmospheric dust or as terrestrial run-off material. They may also reach the water column following resuspension of sediments as a result of wave and currents action or dredging operations. Abiotic particles may be anthropogenic, such as the abundant and ubiquitous microplastics. Particulate matter of biotic origin may be living organisms (see Section 3.3), detritus (nonliving organisms) prior to decomposition, shell fragments, molts, fecal pellets, pieces of tissues dropped during "sloppy feeding," and mucous aggregates formed by zooplankton for feeding (marine snow). Particulate matter in seawater needs to be removed prior to desalination to prevent clogging, in particular in membrane processes. The silt density index (SDI), an on-site measurement of the suspended particles and colloids, is used as an indicator of feedwater quality and as an operational tool for pretreatment adjustments.

3.2 MARINE HABITATS

Habitat is the natural home or physical environment of a living organism. Marine habitats can be divided into coastal and oceanic, and within them into pelagic (water column) and benthic (bottom sediment) habitats (Fig. 3.2). An ecosystem contains the habitat, the organisms living in the habitat, and an energy source, most often sunlight (see Section 3.3).

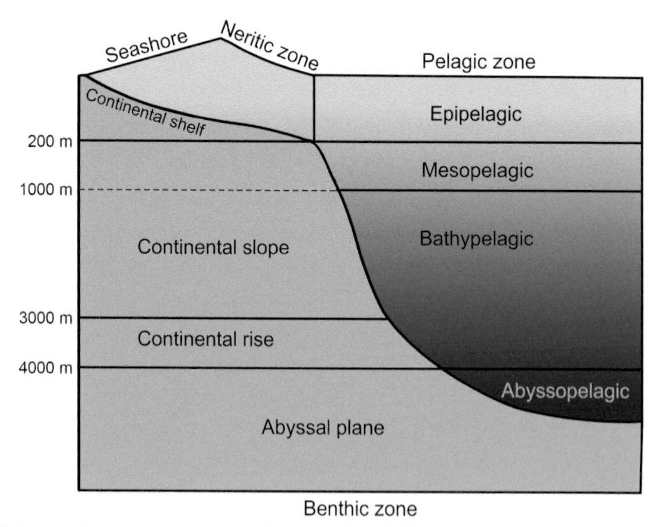

Fig. 3.2 Schematic representation of the coastal and oceanic marine habitats. *In brown,* the benthic zone and *in blue,* the pelagic zone. The depth (*Y*) axis is not to scale.

3.2.1 Coastal Habitat

The coastal habitat encompasses the sea shore, the intertidal zone (the littoral zone subjected to the influence of tides), and the continental shelf, the shallowest part of the continental margin. The continental shelf extends seawards at a gentle bottom slope up 120–200 m depth, about 70 km on average (Fig. 3.2). The neritic pelagic zone includes the water mass located above the continental shelf, almost all within the euphotic zone, where light penetrates and photosynthesis can take place. Only 8% of the ocean's surface area is within the continental shelf, however, biologically, it's the richest area both in the pelagic and benthic compartments. The continental shelf is also the source of seawater for almost all the seawater desalination plants, the area where intakes and also discharge systems are located (see Chapter 4).

3.2.2 Oceanic Habitat

The shelf break, identified by an abrupt increase of bottom slope steepness, is the boundary between the coastal and oceanic habitats. The benthic oceanic habitat includes the continental slope, with a steep slope extending down 2–3 km depth, the continental rise that continues to deepen with a gentle slope, and the deep abyssal plain (deeper than 4 km) (Fig. 3.2). The pelagic domain of the oceanic habitat includes a euphotic zone (down to 200 m water depth), and an aphotic zone, where sunlight does not penetrate. The aphotic zone is subdivided into the mesopelagic (200–1000 m depth), bathypelagic (1000–4000 m depth), and abyssopelagic (deeper than 4000 m depth).

3.3 MARINE ORGANISMS

Biologists classify organisms using the formal taxonomy developed by Carl Linnaeus in 1735. However, other classifications are utilized to simplify or emphasize specific characteristics of organisms such as: size (e.g., picoplankton (0.2–2 μm)); carbon source (autotrophs and heterotrophs); energy source (solar, chemical, respiration); habitat and mobility. For the purpose of this book in seawater desalination, marine organisms are described by habitat and mobility and categorized as plankton, nekton, and benthos.

3.3.1 Plankton

Plankton are organisms that drift with the currents and may have limited self-movement. They are pelagic, living in the water column in the coastal and oceanic habitats. Planktonic organisms are further subdivided into phytoplankton, zooplankton, and bacteria and archea. Within the context of seawater desalination plants, plankton are mostly entrained and entrapped in intake systems while the larger organisms may be also impinged (see Chapter 4).

3.3.1.1 Phytoplankton

Phytoplankton are single cell, eukaryotic (contain a nucleus) microorganisms. They are autotrophic and can produce organic matter from dissolved components in seawater (C, N, and P, primarily) using energy from the sun. Autotrophic organisms are also referred to as photosynthetic organisms and primary producers, being the basis of the food web. Phytoplankton include diatoms, dinoflagellates, and coccolithophores, which are all capable of forming massive blooms. These blooms may be natural, such as the spring bloom or unusual, when anthropogenic nutrients are supplied in excess to the marine environment (Fig. 3.1). These unusual blooms, HABs, increase water turbidity due to high phytoplankton biomass and are popularly known as "red tides" because some bloom species may taint the water red. HAB may consist of toxic and nontoxic species. Toxic species produce toxins that can affect marine organisms directly through exposure or indirectly through consumption of affected organisms. HAB may also produce "foam" (mucilage) that accumulates on beaches and at the water surface. Death followed by bacterial decay of the HAB biomass consumes oxygen dissolved in seawater and may cause hypoxia, anoxia, and mortality of sessile organisms. Desalination plants struggle with HABs due to high particulate content in the feedwater (algal cells and algal organic matter) and the possible presence of toxins (see Section 3.4.1).

3.3.1.2 Zooplankton

Zooplankton are heterotrophic single or multicellular organisms. They cannot synthesize their own food and need to ingest or absorb organic carbon to produce organic matter. Zooplankton range from microscopic organisms, such as protozoa, foraminifera, eggs, and larvae from different species, to large organisms, such as jellyfish and comb jellies. The most abundant zooplankton are the copepod crustaceans.

3.3.1.3 Bacterioplankton and Archaea

Bacterioplankton and archea are single cell, prokaryotic (without a cell nucleus) microorganisms. They may be autotrophic, heterotrophic, or decomposers, the latter breaking down detritus. Bacteria are the most abundant marine organisms and within them *Prochlorococcus marinus*, the most abundant photosynthetic organism on earth. Similar to photosynthesis, certain species of autotrophic bacteria can perform chemosynthesis, using CH_4, H_2S, or metals as an energy source, mainly in dark or oxygen-deficient areas, such as hydrothermal vents.

3.3.2 Nekton

Nekton (or swimmers) are living organisms that are able to swim and move independently of currents. Nekton are heterotrophic and have a large size range, with familiar examples such as fish, squid, octopus, sharks, and marine mammals. Nekton are usually pelagic, living in the water column, but some are demersal and live close to the bottom, both in the coastal and oceanic habitats. Within the context of a seawater desalination plant, nekton may be impinged against the screens of an intake system or entrapped within the system, reducing feedwater withdrawal and impairing the desalination process (see Chapter 4).

3.3.3 Benthos

Benthos or benthic organisms live on the ocean floor, either on the substrate (epifauna and epiflora) or inside it, buried or burrowing in the sediment (infauna). Benthic organisms may be sessile, attached to a firm surface such as rocks and manmade structures, or mobile, moving freely on or in the bottom sediment. Benthos are present in all habitats, from the intertidal to the abyssal plain. Abundance and diversity decrease with increased water depth and distance from the shoreline, a result of dwindling food availability, increased pressure, and decreasing water temperatures. Familiar examples of benthic organisms include macroalgae, seagrasses, corals, barnacles, mussels, sea urchins, and sea stars. Benthos may be: autotrophic, such as seagrasses and algae; heterotrophic, preying on other organisms; filter feeders; feeders on organic matter in the sediment; or decomposers, such as bacteria. Within the context of desalination, benthic organisms may be entrained, impinged, or entrapped in the intake system if the intake head is placed close to the bottom or when organisms, dislodged from the seabed or substrate in events of rough seas, are withdrawn with the feedwater into the intake system.

3.3.4 Invasive Alien Species

This section on marine organisms would not be complete without mentioning invasive alien species (IAS). IAS are nonindigenous organisms, from all taxonomic groups, introduced accidentally or deliberately by human activities into an environment where they are not normally found. They may find an unoccupied environmental niche and thrive, but more often they replace indigenous species with serious negative consequences and changes to the ecosystem. IAS in the marine environment are often introduced by ships (with ballast waters and by attaching to the ship's hull); by mismanagement of marine discharges; as a result of manmade changes to the environment, such as the Suez Canal that connected the Red Sea to the Mediterranean Sea; and from aquaculture, when deliberately introduced nonindigenous species "escape" to the environment.

Within the context of desalination, IAS appearance is unpredictable and may constitute a hindrance to the process, usually by EI&E (see Chapter 4) but also from toxins production in toxic HABs.

3.4 SEAWATER QUALITY AND MARINE ORGANISMS AFFECTING DESALINATION OPERATIONS

The design and planning stages of a desalination plant usually include an environmental impact assessment (EIA) process (see Chapter 7). The EIA identifies the anticipated environmental effects of a proposed plant and recommends measures to minimize them. In addition, the EIA is expected to characterize the marine environment at the vicinity of the plant and to point out environmental characteristics that may influence the desalination process and affect the reliability of freshwater supply. Based on this knowledge, the plant can incorporate measures to minimize possible effects. In the best-case scenario, the marine environment is characterized by repeated marine surveys to determine seawater quality, the habitat and the marine organisms in the area, and also the natural temporal and seasonal variability. However, most often, characterization of the marine environment during an EIA is limited in scope and time and fails to identify factors of concern to plant operations.

The potential of the seawater components and marine organisms to affect desalination operations was described in the previous sections. Below are some actual, real examples, when the quality of seawater hindered and even stopped desalination. Some of the examples were documented in the professional literature or reported in newspapers (Fig. 3.3). However, most temporary plant closures or reduced production go unreported and only personal communication with plant operators can make it known. As a rule, membrane processes are more sensitive to feedwater quality than thermal

Fig. 3.3 News reports on the impact of seawater quality and marine organisms on the operation of desalination plants.

processes. More on the problems facing desalination plants vis-à-vis feedwater (seawater) quality are presented in Chapter 2.

3.4.1 Harmful Algal Blooms

The most studied environmental problem affecting desalination operations are HABs, which have been increasing in frequency worldwide. HABs affect primarily reverse osmosis (RO) plants because the membranes are extremely vulnerable to feedwater quality, making pretreatment exceptionally important. HABs can affect plant operations by clogging fine intake screens (less common) but mainly by affecting or impairing the pretreatment process and therefore failing to protect the RO membranes. As a result, production is reduced or even stopped as pretreatment struggles to remove the increased biomass and to produce the required RO feedwater quality. Furthermore, toxins produced from toxic HABs may pass through the desalination process and enter the product water. A dedicated conference on HAB and desalination took place in 2014 (Muscat, Oman) and was followed by a UNESCO/IOC Report (2017) that included case studies and insights on the pretreatment measures used or tried. Following are some events documented in the report and in other sources.

- The most dramatic and studied HAB event was the 2008–09 extant HAB of the nontoxic marine dinoflagellate *Cochlodinium polykrikoides* in the Gulf, an area with extensive desalination effort. This event caused extensive shutdowns in RO plants, some up to 4 months, due to failure of the pretreatment and was described in the news media and in the professional literature (Fig. 3.3). Short-term shutdowns in two thermal desalination plants were also reported, one associated with foul odor in the product water and the second one to prevent alkaline scaling due to high pH in seawater.
- In March 2011, the La Chimba desalination plant (Antofagasta, Chile) suffered from a HAB of the dinoflagellate *Prorocentrum micans*. The HABs caused severe seawater hypoxia due to the decay of the intense bloom in conjunction with restricted seawater mixing. As a result, H_2S was generated and entered the desalination plant dissolved in the feedwater. However, the plant was able to continue operations due to a warning system and pretreatment measures incorporated following similar previous occurrences.
- From September to November 2009, the Tampa Bay desalination plant (Florida, United States) suffered from recurrent HAB events. In October

2009, the HAB of the dinoflagellate *Ceratium furca* clogged the diatomaceous earth filters of the pretreatment step. The HAB was accompanied by a *Phaeocystis* spp. bloom that caused foaming at the pretreatment basins. Production capacity was reduced.

- In March 2016, the Via Maris desalination plant (Palmachim, Israel) reduced operations due to a HAB of the diatom *A. glacialis*. The HAB was caused by the discharge of poorly treated domestic effluents into the Soreq River that discharges its waters just opposite the intake.
- The Carslbad desalination plant (California, United States) shut down production for 15 days in April 2017, due to a massive HAB at the Agua Hedionda Lagoon. The lagoon is the source of the cooling waters for the power plant adjacent to the desalination plant that in turn feeds the desalination plant. At the same time, along the coast, marine mammals and birds showed symptoms similar to domoic acid (an algal toxin) poisoning.
- In February 2016 and March 2017, a HAB of the dinoflagellate *Noctiluca scintilans* interfered with the desalination in the Gulf.

3.4.2 Marine Organisms

A second unpredictable operating challenge and probably the most graphic is the impingement of marine organism against intake systems reducing or even preventing withdrawal of feedwater into the desalination plant (Figs. 3.3 and 3.4). While EI&E can be minimized during routine operations (see Chapter 4), unusual "blooms" or swarms of organisms challenge the

Fig. 3.4 Jelly-fish impinged against the protective screen of the submerged seawater intake of the Palmachim SWRO, Israel. *Reproduced with permission from Eng. Ofer Fine COO, Palmachim Desalination Plant*

installed measures. "Blinding" of intake screens, in particular by jellyfish but also by seaweeds, macroalgae, comb jellies, and others are widely described in the news media. The emphasis is not only on the operation of conventional and nuclear power plants that utilize seawater as cooling waters but also on desalination plants. Following are some examples.

- From March 2009 to June 2009, the Via Maris desalination plant (Palmachim, Israel) experienced unexplained problems with the pretreatment system, reducing throughput. Examination of the feedwater in the laboratory showed the presence of large numbers of the American comb jelly *Mnemiopsis leidyi* that were clogging the pretreatment. This was also the first record of *M. leidyi*, an IAS introduced to the Black Sea with ballast waters, along the Mediterranean Coast of Israel.
- In 2011, the Orot Rabin power plant and the Hadera desalination plant (Hadera, Israel) were forced to shut down when a swarm of the IAS jellyfish *Rhopilema nomadica* clogged the intakes. A similar event occurred during June 2017.
- In recent years, swarms of jellyfish have caused power plants to shut down in the United States (2012, Diablo Canyon), Sweden (2013, Oskarshamn nuclear power plant), Scotland (2011, Torness power station), Japan (2011, Shimane), and others.
- The Perth Desalination plant (Kawinana, Perth, Australia) reported that intake maintenance has been a challenge over its 10 years of operations. Massive growth of mussels at the intake pipe and screens required yearly cleaning to prevent clogging. Fig. 3.4

3.4.3 Seawater Composition

Changes in seawater composition, a result of pollution events such as polluted riverine runoff and accidental discharges near the intake, were also reported to impair desalination. Following are some documented events.

- In January 1998, the Ajman desalination plant (Abu Dhabi, UAE) was forced to temporarily shut down due to 4000 tons of oil spilled from a barge near Umm Al Quwain.
- In April 2001, the Al Liyya desalination plant in Sharjah and the Al Zawra plant in Ajman (UAE) were closed to prevent contamination by oil pollution origination from a sunken oil/chemical tanker.
- In March 2011, the La Chimba desalination plant (Antofagasta, Chile) experienced high H_2S concentrations in the feedwater due to HABs (see Section 3.4.1). H_2S can readily pass through the RO membranes

and cause foul odor in the product water. However, the plant was able to continue operations due to a warning system and pretreatment measures incorporated following similar previous occurrences.

- In February 2016, the Ashkelon desalination plant (Ashkelon, Israel) was shut down for 4 days due to transboundary pollution in the water, probably domestic sewage, originating at the Gaza Strip.
- During May 2017, the Ashdod, Palmachim, and Soreq desalination plants (Israel) stopped operations due to oil pollution originating from the Ashdod Port.
- On September 2017, the city of Umhlathuze (Western South Africa) temporarily closed the desalination plant due to the presence of manganese in the feedwater and to prevent deposition of manganese salts.

REFERENCES

UNESCO and IOC, 2017. Harmful algal blooms (HABs) and desalination: a guide to impacts, monitoring and management. In: Anderson, D.M., Boerlage, S.F.E., Dixon, M.B. (Eds.), Manuals and Guides.

FURTHER READING

Al-Said, T., Al-Ghunaim, A., Subba Rao, D.V., Al-Yamani, F., Al-Rifaie, K., Al-Baz, A., 2017. Salinity-driven decadal changes in phytoplankton community in the NW Arabian Gulf of Kuwait. Environ. Monit. Assess. 189, 268.

Andrady, A.L., 2011. Microplastics in the marine environment. Mar. Pollut. Bull. 62, 1596–1605.

Bar-Zeev, E., Berman-Frank, I., Liberman, B., Rahav, E., Passow, U., Berman, T., 2009. Transparent exopolymer particles: potential agents for organic fouling and biofilm formation in desalination and water treatment plants. Desalin. Water Treat. 3, 136–142.

Carbonate System. http://cdiac.ess-dive.lbl.gov/ftp/oceans/Handbook_2007/Guide_all_in_one.pdf.

Castro, P., Huber, M., 2007. Marine Biology, sixth ed. McGraw Hill, New York.

Dickson, A.G., Sabine, C.L., Christian, J.R. (Eds.), 2007. In: Guide to Best Practices for Ocean CO_2 Measurements, 3, PICES Special Publication, p. 191.

Elements Present in Seawater. http://www.mbari.org/science/upper-ocean-systems/chemical-sensor-group/periodic-table-of-elements-in-the-ocean.

Galil, B., Kress, N., Shiganova, T.A., 2009. First record of Mnemiopsis leidyi a. Agassiz, 1865 (Ctenophora; Lobata; Mnemiidae) off the Mediterranean coast of Israel. Aquat. Invasions 4, 356–362.

Gases in Seawater. http://www.waterencyclopedia.com/Re-St/Sea-Water-Gases-in.html.

Gross, M.G., Gross, E., 1996. Oceanography. A View of the Earth, seventh ed. Prentice Hall, New Jersey.

Guidelines for Drinking Water Quality. http://www.who.int/water_sanitation_health/publications/2011/dwq_guidelines/en/.

Günel, G., 2016. The Infinity of Water: Climate Change Adaptation in the Arabian Peninsula. Public Culture. Duke University Press.

Isao, K., Hara, S., Terauchi, K., Kogure, K., 1990. Role of sub-micrometre particles in the ocean. Nature 345, 242–244.

Jellyfish and Power Stations. https://www.youtube.com/watch?v=r2YhBc8haPg.

Millero, F.J., 1993. What is PSU? Oceanography 6, 67.

Odum, E.P., 1969. The strategy of ecosystem development. Science 164.

Passow, U., 2002. Transparent exopolymer particles (TEP) in aquatic environments. Prog. Oceanogr. 55, 287–333.

Pilson, M.E.Q., 2013. An Introduction to the Chemistry of the Sea, second ed. Cambridge University Press.

Stockholm Convention on Persistent Organic Pollutants. http://chm.pops.int/Home/tabid/2121/Default.aspx.

Weiss, R.F., 1970. The solubility of nitrogen, oxygen and argon in water and seawater. Deep-Sea Res. Oceanogr. Abstr. 17, 721–735.

Weiss, R.F., 1974. Carbon dioxide in water and seawater: The solubility of a non-ideal gas. Mar. Chem. 2, 203–215.

CHAPTER 4

Theoretical Analysis of the Potential Impacts of Desalination on the Marine Environment

The potential impacts of desalination on the marine environment during plant operations stem from the withdrawal of feedwater (seawater) into the desalination plant and from the discharge of brine from the desalination plant into the marine environment. In this chapter, the physical structures and methods to withdraw feedwater (intake systems) and to discharge brine (discharge systems) are presented, along with their associated potential environmental impacts and possible mitigation measures. Actual observed and documented environmental impacts are presented extensively in Chapters 5 and 6.

4.1 INTAKE SYSTEMS

A robust feedwater intake system is crucial for the reliable operation of a desalination plant: the volume of feedwater supplied by the intake must meet the operational capacity requirements of the plant, and its quality should be consistent. Intake systems are classified into two general categories: (1) open water intakes, consisting of surface open intakes and submerged intakes, located above the seafloor surface, and (2) subsurface intakes, consisting of wells and infiltration galleries, located below the seafloor or below the beach sediments.

The decision as to which intake system to implement is site-specific depending on the natural local hydro-geomorphological conditions, desalination technology, plant size, costs, and environmental considerations. Fig. 4.1 shows a schematic representation of the various intake systems in use by seawater desalination plants.

Marine Impacts of Seawater Desalination: Science, Management, and Policy
https://doi.org/10.1016/B978-0-12-811953-2.00004-9

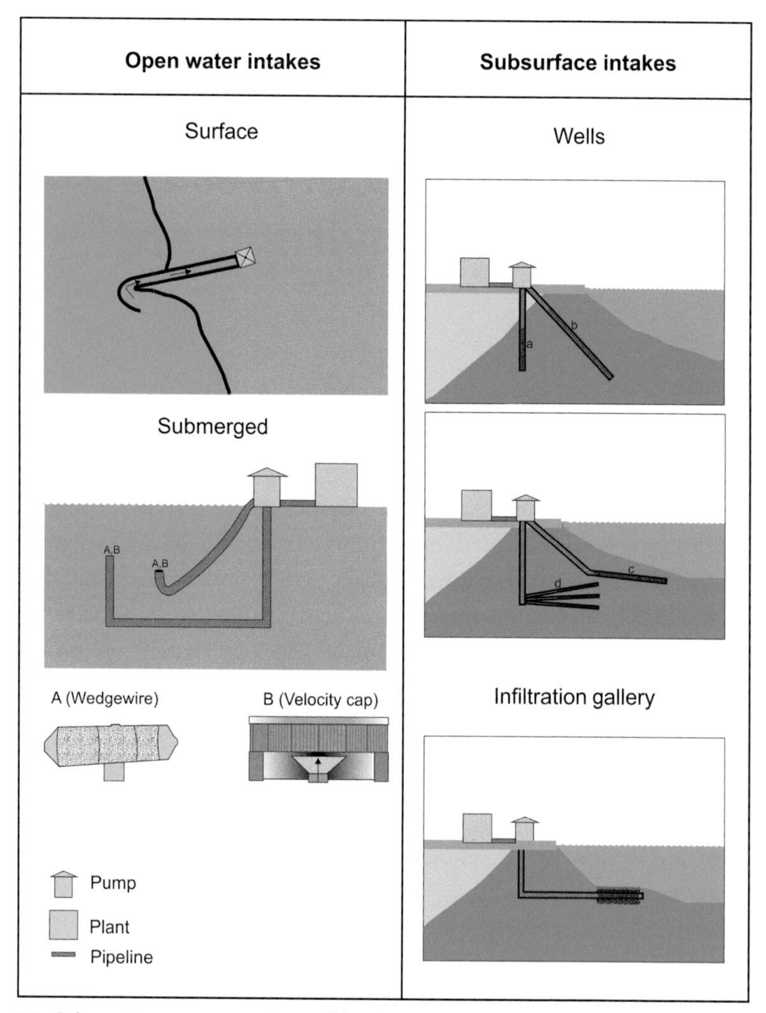

Fig. 4.1 Schematic representation of intake systems. Open water intakes withdraw seawater above the seafloor while subsurface intakes withdraw seawater or brackish water from below the seafloor or beach sediments. Upper right panel: (a) vertical well, (b) slant well. Mid right panel: (c) horizontal well, (d) radial well.

4.1.1 Open Water Intakes

Open water intakes can be located nearshore or offshore. Nearshore intakes, called surface open intakes, withdraw feedwater directly from the sea surface or from shallow waters. Offshore intakes, called submerged open intakes, are located away from the shoreline, and withdraw feedwater from 2 to 6m above the seafloor. Desalination plants co-located with power plants can have a common open intake system or utilize the cooling waters from the power plant as feedwater for desalination. Most thermal desalination plants and large-capacity seawater reverse osmosis (SWRO) desalination facilities use open water intake systems.

4.1.1.1 Surface Open Intakes

Surface open intake systems provide surface and near-surface seawater to the desalination plant. These systems can include dredged channels, going from the surf zone to the plant, or surface water withdrawn from natural or man-made protected, quiescent areas, such as bays, enclosed lagoons, break waters, and stilling basins. Both schemes are followed by a pumping system (Fig. 4.1). As mentioned, desalination plants co-located with power plants can utilize the power plant cooling waters (PPCW) as feedwater through a surface open intake system.

Surface open intakes are usually equipped with active screen systems located onshore to prevent organisms, particles, and debris from entering the plant with the feedwater. An active screen system consists of wire mesh panels mounted in frames of different shapes and incorporates movement, such as horizontal travel and rotation, to remove accumulated debris from the screens. The mesh size of the panels vary, usually 9.5–13 mm, but smaller mesh size are also utilized at specific locations or seasons. A high-pressure spray is commonly used to further dislodge and wash out the accumulated debris from the panels into a trench for subsequent disposal in a landfill or back into the ocean. Some active screens, such as Ristroph traveling screen, are designed to collect organisms and return them unharmed to the source water body by a sluiceway or pipeline. The active screens may be preceded by coarse stationary bar screens (<230 mm), by prescreens (20–150 mm) or by barrier nets extended around the intake zone.

4.1.1.2 Submerged Open Intakes

Submerged open intakes systems are located offshore. They withdraw feed-water from 2 to 6 m above the seafloor and convey seawater to the plant through a pipeline connected to a pump station onshore. At the intake site, the pipeline is equipped with a vertical raiser and an intake head (Fig. 4.1). The pipeline can be laid directly on the seabed, anchored with concrete collars, or laid on a dredged trench and covered with natural sediments. Part of the pipeline, the one closer to the shore, may be laid below the sediment surface using a pipe jacking technique that does not require dredging. As with the surface open intakes, submerged open intakes require seawater screening to prevent organisms, particles, and debris from entering the plant. Screening can be performed with passive screen systems, with no moving parts, installed at the intake head. Screening may be fine (0.5–10.0 mm), such as with a wedgewire, or coarse (50–300 mm) followed by an active screen system onshore. Alternatively, the intake head may be equipped with coarse screen bars and a velocity cap followed by an active screen system onshore. A velocity cap is a flat horizontal cover installed slightly above the terminus of the vertical riser, used to convert vertical to horizontal flow at the intake's head. Velocity caps are installed in most open submerged intakes supplying seawater to large- and mega-sized desalination plants.

4.1.2 Subsurface Intakes

Subsurface intakes (wells and infiltration galleries) are structures built to collect feedwater either from below the beach sediments onshore or from below marine sediments offshore. Favorable hydro-geologic conditions are required for subsurface intakes: strata with high permeability and porosity, such as limestone and alluvial deposits; large areal extent and thickness of the production zone; hydraulic continuity with the ocean; and high recharge rate of the aquifer. The desalination process and plant capacity are additional factors to consider when planning the use of these systems. Subsurface intakes use the sediment as a natural prefiltration step, improving feedwater quality and reducing pretreatment at the desalination plant. Care should be taken when feedwater is anoxic or with high carbonate, iron, or manganese concentrations for they reduce the feedwater quality and may affect the desalination process.

4.1.2.1 Wells

A well is defined as a bored, drilled, or driven shaft whose depth is greater than its largest surface dimension and is used to reach seawater below the sediments, usually at the beach but also offshore. Wells can be vertical, directionally drilled (slant or horizontal) or radial (Ranney collectors) (Fig. 4.1). All include a borehole, a well casing, screens, filter packs, pipelines, and pumps to convey seawater to the plant. The most established technology is the vertical well, constructed near the shoreline as a vertical borehole down to the production zone that is hydraulically connected to the adjoining sea. Vertical wells are used globally as intake systems to hundreds of medium to low capacity SWRO systems. Although vertical wells have a small surface footprint, several wells may be required to supply enough feedwater for desalination. Directionally drilled and radial wells are now under development to be used as intake systems for desalination. Compared to vertical wells, directional and radial wells may supply larger volumes of feedwater due to larger surface area in contact with the groundwater. Moreover, multiple wells can be drilled from a single shaft, reducing the surface footprint compared to multiple vertical well systems.

4.1.2.2 Infiltration Galleries

Infiltration galleries, in contrast to wells, have shallow depths relative to their largest surface dimension. They are essentially in situ sand filters constructed on the beach near or above the high tide line, within the intertidal zone or at the offshore. Galleries consist of excavated trenches, 2–4 m below the sediment surface, with dimensions that are site specific. Galleries are filled with a layer of porous sand, a layer of gravel support, and a screened underdrain and topped with natural local sediments in contact with the seawater. The galleries' walls may be stabilized with geotextile fabrics. Seawater filters through the seabed and the gallery collects at the underdrain and is conveyed to the shore via a pipeline connected to a pump (Fig. 4.1). Galleries as intake systems for desalination plants are under development and in principle may be able to supply feedwater to medium to large capacity plants. Currently they are implemented by a few small SWRO plants. While galleries are more flexible than wells and less dependent on hydro-geological conditions, they are difficult to construct and maintain.

4.1.3 Potential Environmental Impacts of Feedwater Intake Systems

The main potential impacts associated with seawater used as feedwater for desalination are entrainment, impingement, and entrapment (EI&E) of living marine organisms. Additional impacts can originate from the disposal of dead biomass and debris from EI&E and from pipelines and other permanent structures introduced to the marine environment. The magnitude of the effect will depend on the plant's characteristics (intake system and capacity) and on local features at the intake's siting (hydrography, natural population, and environmental status). A desalination intake system located at a site already impacted by other stressors may have a synergistic effect, exacerbating the effects.

4.1.3.1 Effect of Entrainment, Impingement, and Entrapment

EI&E impacts are mainly associated with open intake systems. Subsurface intakes have no direct contact with organisms, conveying seawater from below the sediment to the desalination plant and can essentially be considered not to cause EI&E. EI&E are not constant but subject to daily, seasonal, and temporal variations.

Entrainment is the transport of organisms by the flow of seawater through the intake's screens and into the desalination plant, removing them from marine environment. Entrained organisms are small, typically smaller than 14 mm, with no or limited swimming capability such as bacteria, phytoplankton, algal propagules, eggs, and larvae. Entrainment is generally considered to be proportional to the flow and to cause complete mortality to the entrained organisms.

Impingement refers to the pinning of organisms against the open intake's screens by the velocity and force of the seawater flowing through them into the desalination plant. Impinged organisms may be killed, injured, or weakened, depending on the species, age, and life stage. The potential effects of impingement are loss of abundance and diversity as well as changes in the relative contribution of organisms to the natural community. Impinged organisms are larger than the screen mesh size, typically larger than 14 mm. The magnitude of impingement depends on the flow rate and on the ability of the organism to swim away. Screen cleaning with water jets may also impinge organisms. Impingement can reduce or interfere with the feedwater flow and subsequently with the desalination process (see Chapter 3).

Entrapment occurs when organisms that have entered the intake system are trapped and cannot escape back to their natural habitat. These organisms encompass a large size range, depending on the size of the bar racks and mesh screens installed at the intake head. Entrapped organisms may perish or become resident within the intake system either in the seawater or as colonies, causing biofouling (see Chapters 2 and 3). Colonies include biofilms formed by microorganisms and sessile macro-fouling communities such as algae, bryozoan, barnacles, and mollusks. Entrapped organisms feed on other entrapped or entrained species and on nutrients and organic matter present in the seawater. Except for the case of mortality, entrapment does not reduce abundance but changes the location the species inhabit and may give a developmental advantage to specific species. The sessile communities can reduce or interfere with the feedwater flow and need to be removed periodically (see Section 4.1.3.3).

Assessment of biomass losses due to EI&E. Biomass and biodiversity losses through EI&E are generally addressed qualitatively with almost no quantitative assessments. In principle, hydrodynamic and environmental models can be used to estimate those losses and their environmental impact on ecosystem functions and services (see Chapter 7). However, while hydrographical data may exist for the intake sites, the basic biological data needed for a robust quantitative modeling are lacking globally, even in areas with high desalination efforts. The data needed include: identity, numbers, and biomass of organisms that may be EI&E at the intake's area and at the area hydraulically influenced by it; species characteristics (reproductive cycle, spatial and temporal distributions, habitat and feeding preferences, life span, size); seasonal and temporal variations; and survival rate after EI&E. These biological data are hard, onerous, and expensive to collect; therefore, they are estimated using biological models (see Chapter 7).

4.1.3.2 Effects on Local Aquifers

Withdrawal of feedwater by beach wells may change the direction of the groundwater flow and cause freshwater intrusion in the coastal aquifer. Leakage of feedwater from pipelines to the desalination plant may introduce salt into groundwater and surrounding soils.

4.1.3.3 Effects of Removal of Biomass and Debris

EI&E and debris accumulation in bar racks, screens, and inside the intake system may impede the flow of feedwater and interfere with the desalination process and are therefore removed periodically. Mechanical cleaning may be performed with air bubble bursts, manual scraping, and pigging techniques. When the cleaning is performed onshore (i.e., active screens), the debris and biomass may be disposed of in a landfill with no deleterious effects to the marine environment. When the cleaning of the intake systems returns the debris and biomass to the marine environment, such as in situ cleaning of passive screens or pigging, an array of impacts on the marine environment can be envisaged:

- Aesthetic impact due to the presence of debris and dead organisms.
- Bacterial decomposition of organic biomass and reduction of oxygen concentrations in seawater.
- Mortality of sessile organisms due to hypoxia or anoxia.
- Physical influence of debris on movement and feeding of living organisms.
- Mortality of organisms due to entanglement with debris.

4.1.3.4 Effect on the Benthic Communities

The construction of the intake systems, such as laying out the pipeline, erecting the intake heads, drilling wells, and installing infiltration galleries may have a great impact on the benthic and pelagic communities during the plant's construction stage (see Chapter 7). The physical presence of permanent structures, such as intake heads, passive screens, velocity caps, and pipelines may have the following impact during plant operations:

- Change local ambient currents and water circulation.
- Change sediment transport pattern and sediment characteristics.
- Cause erosion and change bottom bathymetry.
- Provide a hard substrate for biofilm formation and settlement of sessile organisms, endemic and alien.
- Attract free swimming/mobile species to the area.
- All of the above may change the structure of the natural benthic communities and to a lesser degree the pelagic communities, possibly affecting ecosystem functions and services.

4.1.4 Mitigation Measures for the Potential Environmental Impacts of Intake Systems

4.1.4.1 Mitigation of EI&E

Essentially, EI&E may be prevented by using subsurface intake systems. However, subsurface intakes are feasible only in environments with optimal hydro-geological conditions, cannot currently provide enough feedwater for large desalination plants, and may be costly to build and maintain. In open surface intake systems, EI&E may be reduced in the following ways:

- Placing the intake head away from biologically productive areas, such as in deeper, less productive offshore waters.
- Placing the intake head away from marine protected areas and areas with endangered species.
- Improving the recovery of the desalination process, thus reducing the volume of feedwater and EI&E.
- Utilizing the cooling waters of a co-located power plant as feedwater for the desalination plant, preventing additional EI&E.

In addition, impingement and entrapment alone can be mitigated by:

- Physical barriers, such as nets to prevent organisms from reaching the intake area.
- Bypass systems to return unharmed impinged organisms to the natural environment.

- Reduction of feedwater flow velocity (intake velocity slower than 0.15 m/s was found to allow fish to swim away to avoid impingement and entrapment).
- Behavioral barriers such as a velocity cap to convert vertical to horizontal water flow at the intake head (it is known that fish avoid areas with rapid changes in horizontal flow, and this technology is widely applied globally).
- Other behavioral barriers such as sound generating devices, lights, and bubble curtains (their efficacy in preventing impingement is unknown).

Entrainment is harder to mitigate than impingement and entrapment. The use of small mesh size screens may reduce entrainment but will increase impingement. Also, there is a limit as to how small a mesh size can be before it starts to clog and reduce feedwater supply to the plant. As mentioned, positioning the intake head at less productive areas with a smaller population or using subsurface intakes will reduce entrainment.

4.1.4.2 Mitigation of Effects in Local Aquifers

Aquifer overdraft and freshwater withdrawal from the landward direction may be prevented by a rigorous, preconstruction study on the feasibility and placement of the wells.

4.1.4.3 Mitigation of Effects From the Removal of Debris and Biomass

The effects associated with the marine disposal of debris and biomass on the marine environment may be mitigated by reducing impingement, as explained in Section 4.1.4.1, and biofouling. Biofouling may be reduced by:

- Placing the intake away from productive areas.
- Using fouling preventing metal alloys, such as Cu/Ni, for bar racks and mesh screens.
- Using antifouling coating on hard surfaces.
- Applying biocides, such as chlorine, at the intake. However, this measure can affect also the natural populations and react with components in seawater to form toxic chemical species (see Section 4.2.3.1).
- Installing intakes at places with favorable hydrography, such as waves and ocean currents, to reduce settlement.

4.1.4.4 Mitigation of the Impact on the Benthic Communities

The effect of permanent structures on the benthic communities may be minimized by reducing the environmental footprint of the structure (i.e., laying pipelines in covered trenches and not on top of the sediment) and preventing erosion by regular maintenance.

4.2 OUTFALL SYSTEMS

Brine produced by a desalination plant can be disposed of or managed in different ways. Brine can be injected into wells, discharged into inland open waters, discharged into waste water treatment plants, evaporated in drying ponds, used for irrigation and in hybrid systems, or disposed of at sea (see Chapter 2). This section addresses only marine discharge, the system most commonly used by seawater desalination plants. Outfall systems are classified into two general categories: (1) open systems, located at the shoreline or the nearshore, and (2) submerged systems, located at the offshore. Both outfall systems aim to optimize dilution and dispersion of the brine to mitigate potential environmental impacts. Fig. 4.2 depicts a schematic representation of the different outfall systems.

Similar to intake systems, the decision as to which outfall system to implement is site-specific, depending on the local hydrography, desalination technology, plant size, costs, and environmental considerations. In addition, the salinity and temperature of the brine, and hence its density and buoyancy, are important considerations in the decision-making process. As explained in Chapter 2, brine from SWRO plants has about twice the salinity of ambient seawater, is denser than seawater and negatively buoyant, and will sink and disperse close to the bottom as a density current. Brine from thermal desalination plants is saltier (1.2–1.5 times) and warmer (5–15°C) than the ambient seawater, usually lighter than seawater and positively buoyant, and will disperse near the surface. The buoyancy of brine co-discharged with additional effluents is determined by the physical properties of the joint discharge. In the case of a neutrally buoyant discharge, the brine plume will disperse according to the hydrography and water masses of the receiving water body.

4.2.1 Open Outfall Systems

Open discharge systems are located at the shoreline where brine is discharged as a surface stream through open channels (Fig. 4.2). Open systems have traditionally been installed due to their low construction and operational costs compared to submerged systems. Brine in open systems can be diluted with other effluents, either prior or following its discharge at the nearshore. Co-disposed effluents can be the PPCW, treated effluents

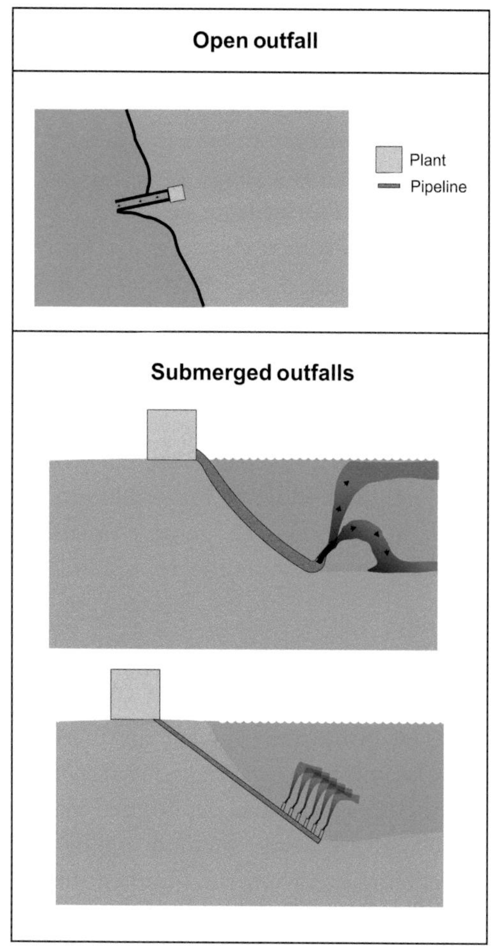

Fig. 4.2 Schematic representation of brine discharge systems. The lower panel shows a submerged outfall equipped with a diffuser system.

from a waste water treatment plant (WWTP), or industrial effluents. The most common and viable option is the co-discharge with PPCW that usually has a flow rate greater than that of the brine. Dilution with WWTP effluent should be discouraged because these effluents may be reused (i.e., for irrigation) after treatment. Brine may be also diluted with seawater prior to discharge, but this option requires additional seawater to be conveyed to the plant, with the concomitant environmental impact. Open discharge systems are commonly used by thermal desalination plants, SWRO co-located with power plants and hybrid thermal-SWRO plants.

4.2.2 Submerged Outfall Systems

Submerged systems, often called marine outfalls, discharge the brine offshore through a pipeline laid on the seabed or in covered trenches, their construction similar to the open submerged intakes (see Section 4.1.1.2). The simplest submerged outfall design is a single-pipe discharge with an inclined open end oriented upwards. The brine jet ascends as it leaves the pipe under pressure and then sinks, ascends, or reaches a neutral buoyancy layer depending on the brine's density (Fig. 4.2). Modern outfall designs incorporate multiport or rosette diffusers to enhance mixing and dispersal of the brine. Diffusers are essentially a series of nozzles through which the brine is discharged, oriented to take advantage of the local currents. Their objective is to increase the mixing of brine with seawater and prevent local accumulation. The optimal location for a marine outfall would be an area with strong currents for quick and effective brine dilution and dispersal, away from biologically productive areas and sensitive or stressed marine ecosystems but close enough to the shoreline to minimize construction and operational costs.

4.2.3 Potential Environmental Impacts of Outfall Systems

The potential effects of brine discharge on the marine environment are directly related to the brine's composition and its dispersion in the receiving environment. Both are site-specific and determined by the desalination process, the plant's capacity and discharge method, and by the hydrography and sensitivity of the receiving environment. As mentioned, the desalination brine is more saline than the receiving environment and also warmer in the case of thermal desalination or co-disposal with PPCW. Brine may contain chemicals used in the pretreatment of the feedwater (coagulants, flocculants, antifoaming, biocides, antiscalants, cleaning agents, corrosion preventers) and post-treatment of product water (pH and mineral adjustors) (see Chapter 2).

The potential impacts of brine discharge on the marine environment may be abiotic or biotic. Abiotic impacts will affect seawater and sediment quality and the natural habitat, while biotic impacts will affect marine organisms. Marine organisms may be further affected by abiotic changes to the environment. Abiotic and biotic impacts will be similar for both outfall systems, with impacts exacerbated for open outfalls due to their proximity to the shoreline, where more productive and sensitive ecosystems are found.

4.2.3.1 Abiotic Impacts

The potential abiotic impacts on the marine environment, categorized by the stressor, are detailed in this section.

Salinity and temperature

- Increase in salinity, or salinity and temperature, of the receiving water body, in particular at the nearshore and in enclosed and semienclosed embayments.
- Changes in water column stratification. Formation of density currents near the bottom when the brine is negatively buoyant and formation of a lighter surficial layer when the brine is positively buoyant.
- Local changes in water circulation and currents due to brine discharge and water column stratification.
- Reduced oxygen solubility in seawater due to higher salinity and temperature.
- Reduced oxygen concentrations due to stratification.
- Increase in salinity in the sediments' pore water when brine disperses near the bottom as density currents.

Chemicals used in the pretreatment and post-treatment stages

- Increase in water turbidity and decreased light penetration due to the discharge of particles, such as coagulants with the backwash and pH adjustors.
- Aesthetic impact due to turbidity and water discoloration.
- Increase in environmental concentrations of chemicals discharged with the brine.
- Accumulation of pollutants in enclosed discharge areas.
- Changes in pH. Expected to be minor due to the natural buffering capacity of seawater.
- Formation of toxic complexes upon mixing of the brine with seawater. For example, formation of bromoform from the reaction of residual chlorine used as biocide with bromide present in seawater.

Metals from corrosion (copper, nickel, iron, chromium, other)

- Increase in environmental concentrations of corrosion material, mainly in the sediments.
- Increased in particles in the water column.

Pipelines and permanent structures

- Introduction of hard substrates to the area.
- Changes in local ambient currents.
- Changes in sediment transport pattern and sediment characteristics.
- Erosion and changes to bottom bathymetry.
- Changes in habitat.

4.2.3.2 Biotic Impacts

Biotic impacts affecting marine organisms are derived directly from the toxicity or deleterious effects of the brine's constituents and indirectly from abiotic changes. The potential biotic effects encompass all levels of the marine ecosystem: from bacteria and microalgae to fish and their predators, in seawater and sediments, and from the onshore to the offshore (see Chapter 3). Overall, biotic impacts may be lethal or sublethal, change biomass and diversity, change the relative contribution of organisms to the natural community, alter behavior and life cycle, and change ecosystem structure, functions, and services.

Since a specific biotic effect can be caused by different components of the brine and by abiotic changes, biotic impacts are categorized and detailed below by effect.

Mortality. Biota mortality can be caused by the following factors:

- Salinity and temperature in excess of organism tolerance inducing a haline or thermal shock.
- Nonneutralized biocides and toxic complexes created by the reaction of brine components with seawater.
- Toxicity of a specific brine component or a mixture of compounds.
- Hypoxia and anoxia, for sessile biota.
- Water turbidity and decreased light penetration, for photosynthetic organisms.
- Changes in sediment composition and benthic habitat, for benthic organisms.

Sublethal effects. Sublethal effects can be caused by the following factors:

- Increased salinity and temperature that may:
  - modify photosynthetic rates in photosynthetic organisms;
  - modify metabolism, physiology and growth in organisms;
  - provide a developmental advantage to thermal and haline tolerant species; and
  - facilitate the introduction of nonindegenous species.
- Changes in water column stratification and currents may alter the transport of planktonic species at the vicinity of the discharge area.
- Hypoxia and anoxia may drive away organisms with swimming capabilities or deter them from arriving.
- Changes in sediment characteristics and transport, and permanent structures affecting the habitat may change the benthic and sessile community structure. They may also promote settlement of nonindigenous species.

- Loss or changes in habitat may impact spawning, feeding grounds and migration, changing the biotic community.
- Chemicals discharged with the brine may have a chronic influence on the biota, affecting metabolism and genetic diversity.
- Nutrients and iron discharge may trigger unnatural phytoplankton blooms.

4.2.4 Mitigation Measures for the Potential Environmental Impacts of Outfall Systems

Potential abiotic and biotic impacts of brine discharge on the marine environment may be mitigated. The following ways, some already explained in Chapter 2, are envisaged:

- Elimination of brine disposal using zero liquid discharge (ZLD) technologies.
- Reduction of brine volume by improving the efficiency of the desalination process.
- Increased brine dilution, to reduce salinity, temperature, and concentration of chemicals present in the brine.
- Siting the outfall system away from productive and ecological sensitive areas.
- Improvement of brine quality by reducing chemical use and discharge, and by using more environmental friendly ("greener") compounds.
- Treatment of brine prior to discharge to remove potentially harmful chemicals.

FURTHER READING

Ahmad, N., Baddour, R.E., 2014. A review of sources, effects, disposal methods, and regulations of brine into marine environments. Ocean Coast. Manag. 87, 1–7.

Ahmed, M., Shayya, W.H., Hoey, D., Al-Handaly, J., 2001. Brine disposal from reverse osmosis desalination plants in Oman and the United Arab Emirates. Desalination 133, 135–147.

Barnthouse, L.W., 2013. Impacts of entrainment and impingement on fish populations: a review of the scientific evidence. Environ. Sci. Pol. 31, 149–156.

Cooley, H., Ajami, N., Heberger, M., 2013. Key issues in seawater desalination in California. Marine impacts. Pacific Institute Report.

Ferry-Graham, L., Dorin, M., Lin, P., 2008. Understanding entrainment at coastal power plants: informing a program to study impacts and their reduction. California Energy Commission, PIER energy-related environmental research program. CEC-500-2007-120.

Foster, M.S., Cailliet, G.M., Callaway, J., Raimondi, P., Steinbeck, J., 2012. Mitigation and fees for the intake of seawater by desalination and power plants. Report to State Water Resources Control Board, Sacramento, California.

Gille, D., 2003. Seawater intakes for desalination plants. Desalination 156, 249–256.
Giwa, A., Dufour, V., Al Marzooqi, F., Al Kaabi, M., Hasan, S.W., 2017. Brine management methods: recent innovations and current status. Desalination 407, 1–23.
Hogan, T.W., 2015. Impingement and entrainment at SWRO desalination facility intakes. In: Missimer, T.M., Jones, B., Maliva, R.G. (Eds.), Intakes and Outfalls for Seawater Reverse-Osmosis Desalination Facilities: Innovations and Environmental Impacts. Springer International Publishing, Cham, pp. 57–78.
Jenkins, S., Paduan, J., Roberts, P., Schlenk, D., Weis, J., 2012. Management of brine discharges to coastal waters. Recommendations of a science advisory panel. Technical Report 694, Southern California Coastal Water Research Project, Costa Mesa, CA.
Kress, N., Galil, B.S., 2016. Impact of seawater desalination by reverse osmosis on the marine environment. In: Burn, S., Gray, S. (Eds.), Efficient Desalination by Reverse Osmosis. IWA, London.
Lattemann, S., Hopner, T., 2008. Impacts of seawater desalination plants on the marine environment of the Gulf. In: Abuzinada, A.H., Barth, H.J., Krupp, F., Böer, B., Al Abdessalaam, T.Z. (Eds.), Protecting the Gulf's Marine Ecosystems from Pollution. Birkhäuser Verlag, Switzerland, pp. 191–205.
Lokiec, F., 2013. Sustainable desalination: environmental approaches. The International Desalination Association World Congress on Desalination and Water Reuse, Tianjin, China.
Mackey, E.D., Pozos, N., Wendle, J., Seacord, T., Hunt, H., Mayer, D.L., 2011. Assessing Seawater Intake Systems for Desalination Plants. Water Research Foundation, Denver, CO.
Maliva, R.G., Missimer, T.M., 2015. Well intake systems for SWRO systems: design and limitations. In: Missimer, T.M., Jones, B., Maliva, R.G. (Eds.), Intakes and Outfalls for Seawater Reverse-Osmosis Desalination Facilities: Innovations and Environmental Impacts. Springer International Publishing, Cham.
Mayhew, D.A., Jensen, L.D., Hanson, D.F., Muessig, P.H., 2000. A comparative review of entrainment survival studies at power plants in estuarine environments. Environ. Sci. Pol. 3 (Suppl. 1), 295–301.
Meneses, M., Pasqualino, J.C., Cespedes-Sanchez, R., Castells, F., 2010. Alternatives for reducing the environmental impact of the main residue from a desalination plant. J. Ind. Ecol. 14, 512–527.
Mezher, T., Fath, H., Abbas, Z., Khaled, A., 2011. Techno-economic assessment and environmental impacts of desalination technologies. Desalination 266, 263–273.
Mickley, M.M., 2006. Membrane concentrate disposal: practices and regulation. Desalination and water purification research and development program No. 123 (second edition). U.S. Department of Interior.
Missimer, T.M., Maliva, R.G., 2018. Environmental issues in seawater reverse osmosis desalination: intakes and outfalls. Desalination 434, 198–215.
Missimer, T.M., Ghaffour, N., Dehwah, A.H.A., Rachman, R., Maliva, R.G., Amy, G., 2013. Subsurface intakes for seawater reverse osmosis facilities: capacity limitation, water quality improvement, and economics. Desalination 322, 37–51.
Missimer, T.M., Hogan, T.W., Pankratz, T., 2015. Passive screen intakes: design, construction, operation, and environmental impacts. In: Missimer, T.M., Jones, B., Maliva, R.G. (Eds.), Intakes and Outfalls for Seawater Reverse-Osmosis Desalination Facilities: Innovations and Environmental Impacts. Springer International Publishing, Cham.
Murray, J.B., Wingard, G.L., 2006. Salinity and temperature tolerance experiments on selected Florida bay mollusks. U.S. Geological Survey Open-File Report 1026, p. 59.
NRC, 2008. Desalination, a National Perspective National Research Council of the National Academies. The National Academies Press, Washington, DC.
Pankratz, T., 2015. Overview of intake systems for seawater reverse osmosis facilities. In: Missimer, T.M., Jones, B., Maliva, R.G. (Eds.), Intakes and Outfalls for Seawater

Reverse-Osmosis Desalination Facilities: Innovations and Environmental Impacts. Springer International Publishing, Cham, pp. 3–17.

Peters, T., Pinto, D., 2008. Seawater intake and pre-treatment/brine discharge—environmental issues. Desalination 221, 576–584.

Purnama, A., 2015. Environmental quality standards for brine discharge from desalination plants. In: Baawain, M., Choudri, B.S., Ahmed, M., Purnama, A. (Eds.), Recent Progress in Desalination, Environmental and Marine Outfall Systems. Springer International Publishing, Cham, pp. 257–267.

Purnama, A., Al-Barwani, H.H., Smith, R., 2005. Calculating the environmental cost of seawater desalination in the Arabian marginal seas. Desalination 185, 79–86.

Strange, E.M., 2011. Methods for evaluating the potential cumulative effects of power plant intakes on coastal biota. California Energy Commission, PIER energy-related environmental research program. CEC-500-2010-028.

UNEP. 2008. Desalination resource and guidance manual for environmental impact assessments. United Nations Environment Programme, Regional Office for West Asia, Manama, and World Health Organization, Regional Office for the Eastern Mediterranean, Cairo Ed. S. Lattemann: 168 pp.

US-EPA, 2014. National pollutant discharge elimination system—Final regulations to establish requirements for cooling water intake structures at existing facilities and amend requirements at phase I facilities. Final rule, Federal Register 79 (158). 15 August 2014.

Villacorte, L.O., Tabatabai, S.A.A., Anderson, D.M., Amy, G.L., Schippers, J.C., Kennedy, M.D., 2015. Seawater reverse osmosis desalination and (harmful) algal blooms. Desalination 360, 61–80.

Voutchkov, N., 2011. Overview of seawater concentrate disposal alternatives. Desalination 273, 205–219.

WHO, 2007. Desalination for safe water supply. Guidance for the health and environmental aspects applicable to desalination. World Health Organization, WHO/SDE/WSH/07.

CHAPTER 5

Early Observations of the Impacts of Seawater Desalination on the Marine Environment: From 1960 to 2000

This chapter and Chapter 6, describing the actual observed impacts of seawater desalination reported in the peer-reviewed and gray literature are at the heart of this book. As it will become clear, the number of reports is limited and their conclusions contradictory at times. However, the number of publications have been increasing with time in pace with the increase in desalination effort and the awareness that desalination poses an environmental price that needs to be understood and alleviated (Fig. 5.1).

In the early literature, from the 1960s to 2000, the impacts were mostly addressed qualitatively (see Chapter 4) with only 17 studies actually describing impacts, if any, on the marine environment. This scant number may be a result of pressing timelines to produce potable water and the notion that if properly engineered, desalination would not affect the environment. These 17 early observations are described in this chapter. Six studies were conducted in the Middle East, three in the Caribbean, three in the Gulf of Mexico, and one in each of the following places: Australia, Channel Islands, Japan, Italy, and Spain (Canary Islands) (Fig. 5.2). Ten studies addressed thermal desalination plants, and seven addressed reverse osmosis plants or salinity as a stressor.

From the start of large-scale desalination and until 2005, the main processes used were thermal (multistage flash distillation (MSF) and multieffect distillation (MED), see Chapter 2). Membrane processes (mainly reverse osmosis (RO)) steadily increased from almost none in the 1960s, through ca. 25% of the total desalination effort in 1980 to about 45% in 2000. Accordingly, this chapter is divided into observed impacts associated with thermal desalination followed by impacts associated with RO desalination.

Marine Impacts of Seawater Desalination: Science, Management, and Policy
https://doi.org/10.1016/B978-0-12-811953-2.00005-0

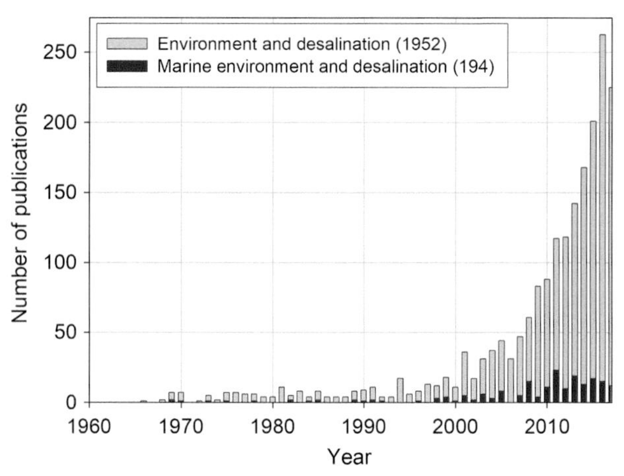

Fig. 5.1 The number of publications addressing environment and desalination (1952 studies, in *gray*) and addressing marine environment and desalination (194 studies, in *red*) from 1960 to 2017. *(Data were collected using Scopus on January 1st, 2018.)*

5.1 IMPACTS ASSOCIATED WITH DESALINATION USING THERMAL PROCESSES

As detailed in Chapter 2, seawater desalination by thermal processes produces brine with temperature and salinity higher than the receiving environment by 5–15°C and 1.2–1.5 times, respectively. The brine may include chemicals used in the desalination process and metals from internal corrosion of pipes, valves, and pumps due to high temperature operations and periodic acid cleaning.

Following are the summaries of the 10 studies identified, in chronological order.

- An extensive field and laboratory study was performed at a large-scale MSF thermal desalination plant in Key West, Florida (Gulf of Mexico), during 18 months in 1969–70. The brine was discharged with temperature and salinity higher than ambient by 0.3–0.4°C and 0.2–0.5, respectively. Upon discharge, the brine sank and accumulated at the bottom, due to the local bathymetry, stratifying the water column and reducing water circulation. The settled brine was thus prevented from dispersing to adjacent natural seagrass environment. The copper concentration in the brine was 5–10 times above ambient, often at levels above toxicity thresholds for native species. A variety of organisms vanished from the discharge site during the study: sea squirts, algae, bryozoans, and sabellidae worms. Sea urchins suffered mortality on the nearby seagrass fields,

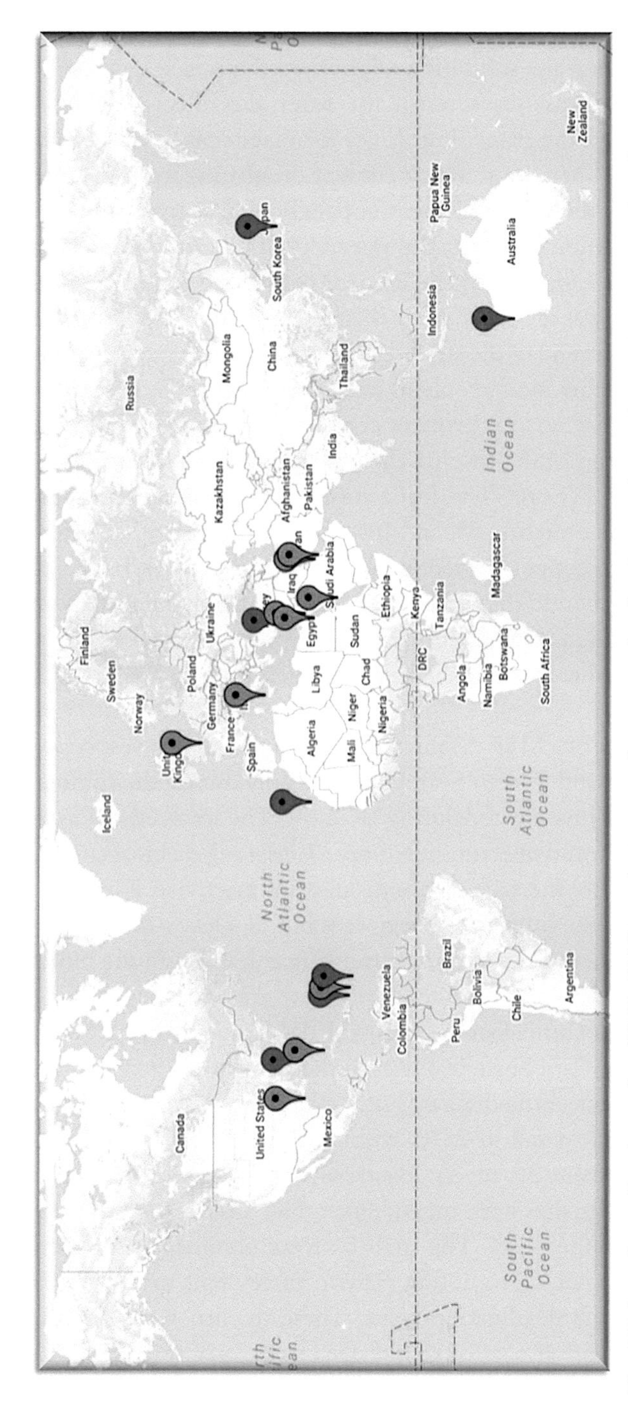

Fig. 5.2 Locations of early studies (1960–2000) on the impact of seawater desalination on the marine environment. *Red* markers represent thermal desalination, and *blue* markers represent reverse osmosis.

probably a result of Cu toxicity as shown in bioassay experiments. Bioassays showed also inhibition of photosynthesis in seagrasses and mortality of Ascidians (sea squirts), the latter also affected by temperature. Periodically, the plant shut down for maintenance or cleaning. When it resumed operations, a dark effluent originating from the maintenance operations, polluted with ionic copper but with temperature and salinity similar to ambient values, did not sink but dispersed to the surrounding areas. The impact from the effluent discharged following maintenance operations was more extensive than the impact of the regular desalination brine. At the end of the study, extensive engineering changes were made to correct corrosion problems and lower copper discharge.

- A 60-day bioassay experiment conducted in 1969–71 tested the impact of brine from MSF plants on the oyster *Crassostrea virginica* from the Gulf of Mexico (Texas). The brine mixture had temperature, salinity, and copper concentration 5% and 10% above the ambient seawater. The relatively high copper content of the brine caused up to 100% mortality of juvenile oysters in the most drastic treatment. Copper bioaccumulated in the adult oysters, and temperature and salinity induced mortality as a result of reduced resilience to a pathogen fungus. Spawning inhibition was observed as well.
- The impact of a MSF desalination plant in the Southwest coast of Jersey, Channel Island (North Sea), operating only during the summer tourism season, was investigated during 1972–74. The seaweeds *Fucus serratus* and *Fucus spiralis* and the common limpet *Patella vulgata* bioaccumulated copper even after the winter months of plant inactivity. Zinc also bioaccumulated in the limpets and seaweeds but to a lesser degree than copper. No differences were found in iron concentrations in the biota from the discharge and reference sites.
- Dispersion of brine from a small MSF desalination plant in St. Croix, Virgin Islands (Caribbean Sea), was studied before 1979. The brine affected a small area near the discharge site: seawater temperature decreased from 34°C at the outfall to 28°C within 12 m of the discharge, and salinity decreased from 36 to 35. Bacterial counts in seawater at the intake and discharge sites were much higher than those found in reference areas away from the plant. The high bacterial abundance was attributed to organic nutrients promoting growth of resilient species in the coolest region of the desalination plant. These in turn were discharged with the brine and also affected the feedwater quality. No similar impact was observed at an RO plant operating for 3 months in the area.

- The impact of the combined discharges of a power and a thermal desalination plant in Eilat, Israel (Red Sea), was studied in the late 1970s. The discharges had high temperatures and salinity compared to the ambient conditions (39°C and 46 salinity compared to 22°C and 40, respectively). The effluent dispersed close to the bottom and was detected even 200 m away from the discharge area. The combined discharges also had high concentrations of copper and iron that precipitated as hydroxides at the vicinity of the outfall. An area 70 m wide opposite the discharge point was almost devoid of biota, while deformed sea urchins were found beyond that. The growth rate of the deformed sea urchins transferred to aquaria was low compared to the growth rate of nonimpacted specimens under the same conditions.
- Brine from the Sitra power and thermal desalination plants in Bahrain (the Gulf, studied before 1990) significantly increased the temperature (from 22°C to up to 37°C) and salinity (from 46 up to 55) of the receiving waters within 70 m from the outfall due to restricted water circulation. No changes in pH were measured. A slight decrease in dissolved oxygen was detected at some areas (from 6.5 to 5.0 mg/L).
- In Jeddah, Kingdom of Saudi Arabia (KSA, Red Sea, study from 1991), a large MSF desalination plant discharged brine with salinity of 47.5 compared to 39 in the receiving waters and temperature higher by 9°C. No dispersion pattern was described.
- The benthic communities along the KSA's Gulf coastline were assessed in 1985–86. The area, stressed by discharges from power plants, thermal desalination plants, and oil spills, consisted of seagrass meadows and of sand-silt habitats. Salinity was found as the most important driver of change: increased salinity decreased the diversity of the benthic communities in both seagrass and sand/silt. Hypersalinity reduced the abundance of the benthic organisms in the seagrass but increased the biomass of those communities in the sand-silt habitat. Although the main factor responsible for increasing salinity—restricted circulation, high evaporation, and/or brine discharge—was unclear, this study exemplifies the impacts of brine discharge.
- Corals and some planktonic organisms disappeared from the coastal areas off Hurghada City, Egypt (Red Sea, study from the early 1990s) due to the discharge of brines from thermal desalination plants. The populations of some fish species declined as well and settlement and establishment of species from other areas were reduced. The warm and saline brine (27.5°C and 60 salinity compared to the ambient 23°C and 43) was

discharged also with a low pH of 5.8 compared to natural seawater pH of 8.3. The extent of the brine dispersion was not given in this study.

- Benthic communities were studied at the discharge site of a thermal compression desalination plant in Ustica Island (Italy, Mediterranean Sea), in 1996, before the plant was activated, and twice in 1997. Brine with a salinity of 70 and 3°C warmer than ambient was discharged at 50 m depth through a marine outfall equipped with diffusers. Temperature at the outfall was higher than ambient by 2–3°C and salinity was 49.6 and 47.4 in summer and winter, respectively, 30% higher than the ambient salinity (37.9 and 36.9, respectively). Dilution to ambient values occurred within a few meters from the discharge. Prior to the activation of the plant, polychaeta formed the most abundant group followed by crustaceans. The sipunculid (marine worm) *Aspidosiphon muelleri* was the dominant species in the area and the polychaeta *Lysidice ninetta* was well-represented. Brine discharge reduced the general diversity and abundance at the outfall, in particular of crustaceans, mollusks, and echinoderms that nearly disappeared. No impact was detected 25 m from the outfall. Specifically, the dominant *L. ninetta* did not seem to be affected by the brine discharge while *A. muelleri* was not found at the outfall. In the winter, the polychaeta *Pisione remota* dominated the outfall station and was found only there.

5.2 IMPACTS ASSOCIATED WITH REVERSE OSMOSIS DESALINATION

As detailed in Chapter 2, seawater desalination by RO produces brine with salinity about twice the salinity of seawater and with temperatures near the ambient temperatures. The brine may include also chemicals used in the desalination process, in particular coagulants and antiscalants. Only seven early studies were identified, among them three showing the impact of increased salinity not linked to desalination but relevant as a proxy to hypersalinity impact. Following are the summaries of these studies, first the three studies conducted within the context of desalination, in chronological order, followed by the three proxy studies.

- A bioassay study was conducted in Japan in 1991 to elucidate the impact of brine discharged from a desalination plant on coastal marine organisms. Fish and bivalves (fertilized eggs, larvae, and juveniles) were exposed to salinities ranging from 30 to 100. Mortality of sea bream juveniles, flounder larvae, and soft bivalve started at 45, 50, and 50 salinities, respectively, while hatchability of flounder stopped at salinity of 40. The critical salinity for coral survival was 43.

- An in situ experiment was conducted for 6.5 months in Antigua, West Indies (Caribbean Sea), during 1997, where brine from an RO desalination plant (salinity of 45 at the discharge area) was diverted into a seagrass meadow. Brine did not affect shoot density, areal blade biomass and productivity of the turtle grass (*Thalassia testudinum*) nor the grazing rate of the bucktooth parrotfish (*Sparisoma radians*). Moreover, no impact was detected on the biomass of benthic microalgae nor on the abundance of benthic diatom and foraminifera. No obvious stress or mortality was observed in the following organisms: the sessile soft coral *Pseudopterogorgia acerosa*, the hard corals *Porites astreoides* and *Porites furcata*, the mobile queen conch *Strombus gigas*, and the cushion starfish *Oreaster reticulatus*. The only impact detected during the experiment was an increase in biomass of the brown macroalgae *Dictyota dichotoma* after 3 months at the brine discharge, attributed to episodic nitrogen supply either from filter backwash, detergents used in the desalination, or stormwater runoff from the desalination plant.
- The Dhekalia SWRO desalination plant in Cyprus (Mediterranean Sea) started to operate in 1997. Results of monitoring studies, reported after 3 years of operation, state that brine discharge had negligible impact, limited to an area with radius of 200 m from the outfall. Some changes of the benthic community in the discharge area were observed right after the plant started to operate, such as the disappearance of sea urchins, but no additional changes were detected thereafter. However, the study does not describe the impacts or changes that occurred in the discharge area.
- Brine from the Maspalomas II SWRO desalination plant (Canary Islands, Spain, Atlantic Ocean) was discharged through two outfalls located at 7.5 m water depth. The brine had a salinity of 75, and its temperature was 2°C higher than ambient. During the survey in August 2000, seawater salinity was 38.44 at 20 m from the discharge point and similar to ambient (S = 37) 100 m away. The dilution was fast and across the whole water column due to the shallow depth and dynamic hydrography in the area. Brine discharge did not affect seawater turbidity. Brine did not affect the "cebadales," an association of two seagrasses: *Cymodea nodosa* and *Caulerpa prolifera*. They looked healthy with a rich accompanying biological community, in particular young fish.
- Brine discharged from salt works into the Bahia Fosforescente, Puerto Rico (Caribbean Sea), in 1969 increased bottom salinity from 37 to

57 causing stagnation, anoxia, and H_2S formation. This hypersaline water remained trapped in the bay for 3 weeks until strong winds mixed it with ambient waters and restored the previous conditions. The bottom brine caused phytoplankton mortality, accumulated pheophytin, a decomposition product of chlorophyll, and turned a yellow-green color.

- An inkling to the salinity tolerance of seagrasses can be exemplified by a study from 1983 on the adaptation of *Amphibolis antarctica* to the natural high salinity range in Shark Bay, Western Australia (Indian Ocean). The naturally occurring salinity in the bay ranged from 35 to 65. Maximal leaf production rates were obtained at 42 salinity. Seedlings showed marked senescence within 5 days of being placed in a salinity of 65. The results of the experiment are consistent with the suggestion that the observed in situ decline in seagrass biomass, area coverage, and productivity were a result of exposure to high salinity.
- Colonies of the stony coral *Siderastrea siderea* were exposed to salinities lower and higher than the ambient salinity of 28–30 off the coast of Panacea, Florida (Gulf of Mexico), in a study conducted before 1987. *S. siderea* was able to acclimate to a salinity of 42 when the salinity was increased slowly over a 30-day period. Sudden salinity changes proved fatal to the coral. After acclimation, neither respiratory nor photosynthetic rates were affected by changes in salinity of less than 10 above or below the acclimation salinity. Larger changes in salinity (either up or down) decreased respiratory and photosynthetic rates proportionally to the magnitude of the salinity change.

5.3 INTEGRATION OF THE EARLY RESULTS

The most evident impacts of brine discharge from thermal desalination plants, as described is the early studies, were increases in temperature and salinity of the receiving environment, their magnitude being site specific. However, even the use of these simple parameters to follow the dispersion of the brine in situ was seldom reported. In addition, at the earlier stages of thermal desalination, scaling was reduced mainly by the use of acids that in turn promoted corrosion and increased discharge of metals (in particular copper, zinc, and iron) and low-pH effluents to the marine environment. Most, but not all, of the observed chemical-biological impacts, were attributed to metals from corrosion. Those included deposition of metal hydroxides on the sediments, bioaccumulation by biota, reduction in biotic abundance and diversity, mortality, deformation, and changes in metabolic processes.

Dispersion of brine from RO desalination plants in the marine environment was reported only in one of the early studies. Brine did not affect the biota in two in situ studies while in a third a limited biotic effect was reported. Bioassays, studying the effect of hypersalinity on biota, found effects starting at a salinity of 40, highly dependent on the species tested and its developmental stage.

FURTHER READING

Al-Mutaz, I.S., 1991. Environmental impact of seawater desalination plants. Environ. Monit. Assess. 16, 75–84.

Altayaran, A.M., Madany, I.M., 1992. Impact of a desalination plant on the physical and chemical properties of seawater, Bahrain. Water Res. 26, 435–441.

Argyrou, M., 1999. Impact of Desalination Plant on Marine Macrobenthos in the Coastal Waters of Dhekelia Bay, Cyprus. Department of Fisheries, Ministry of Agriculture, Natural Resources and Environment, Cyprus.

Castriota, L., Beltrano, A.M., Giambalvo, O., Vivona, P., Sunseri, G., 2001. A one-year study of the effects of a hyperhaline discharge from a desalination plant on the zoobenthic communities in the Ustica Island marine reserve (southern Tyrrhenian Sea). Rapport Commission Internationale pour l'Exploration Scientifique de la. Méditerranée 36, 369.

Chesher, R., 1971. Biological Impact of a Large-Scale Desalination Plant at Key West, Florida. Elsevier Oceanography Series, vol. 2. pp. 99–164.

Chesher, R.H., 1975. Biological impact of a large-scale desalination Plant at key West, Florida. In: Wood, E.J.F., Johannes, R.E. (Eds.), Elsevier Oceanography Series. Elsevier, pp. 99–153 (Chapter 6).

Cintrón, G., Maddux, W.S., Burkholder, P.R., 1970. Some consequences of brine pollution in the Bahía Fosforescente, Puerto Rico. Limnol. Oceanogr. 15, 246–249.

Coles, S.L., McCain, J.C., 1990. Environmental factors affecting benthic infaunal communities of the Western Arabian gulf. Mar. Environ. Res. 29, 289–315.

Dafni, J., 1980. Abnormal growth patterns in the sea urchin Tripneustes CF. Gratilla (L.) under pollution (Echinodermata, Echinoidea). J. Exp. Mar. Biol. Ecol. 47, 259–279.

Hammond, M., Blake, N., Hallock-Muller, P., Luther, M., Tomasko, D., Vargo, G., 1998. Effects of disposal of seawater desalination discharges on near Shore Benthic Communities. Report of Southwest Florida Water Management District and University of South Florida.

Iso, S., Suizu, S., Maejima, A., 1994. The lethal effect of hypertonic solutions and avoidance of marine organisms in relation to discharged brine from a destination plant. Desalination 97, 389–399.

Mabrook, B., 1994. Environmental impact of waste brine disposal of desalination plants, Red Sea, Egypt. Desalination 97, 453–465.

Mandelli, E.F., 1975. The effects of desalination brines on Crassostrea virginica (Gmelin). Water Res. 9, 287–295.

Muthiga, N.A., Szmant, A.M., 1987. The effects of salinity stress on the rates of aerobic respiration and photosynthesis in the hermatypic coral Siderastrea Siderea. Biol. Bull. 173, 539–551.

Pérez Talavera, J., Quesada Ruiz, J., 2001. Identification of the mixing processes in brine discharges carried out in Barranco del Toro Beach, south of gran Canaria (Canary Islands). Desalination 139, 277–286.
Romeril, M.G., 1977. Heavy metal accumulation in the vicinity of a desalination plant. Mar. Pollut. Bull. 8, 84–87.
Tomasko, D.A., Blake, N.J., Dye, C.W., Hammond, M.D., 1999. Effects of the disposal of reverse osmosis seawater desalination discharges on a seagrass meadow (Thalassia testudinum) offshore of Antigua, West Indies. In: Bortone, S.A. (Ed.), Seagrasses: Monitoring, Ecology, Physiology, and Management. CRC Press, pp. 99–112.
Tsiourtis, N.X., 2001. Seawater desalination projects. The Cyprus experience. Desalination 139, 139–147.
Walker, D.I., McComb, A.J., 1990. Salinity response of the seagrass Amphibolis Antarctica (Labill.) Sonder et Aschers.: An experimental validation of field results. Aquat. Bot. 36, 359–366.
Winters, H., Isquith, I.R., Bakish, R., 1979. Influence of desalination effluents on marine ecosystems. Desalination 30, 403–410.

CHAPTER 6

Actual Impacts of Seawater Desalination on the Marine Environment Reported Since 2001

The early observations on the impacts of seawater desalination on the marine environment were described in Chapter 5. In this chapter, a detailed account of actual studies reported since 2001 is presented followed by a critical analysis of the reports. As in Chapter 5, this chapter addresses only the effects of brine discharge. There are no actual studies on intake effects, except for a few studies performed in power plants concerning feedwater withdrawal for cooling and its effect on entrainment, impingement, and entrapment (EI&E) (see Chapter 4).

The chapter includes descriptions of in situ studies, in situ experiments, and laboratory bioassays performed on local natural populations exposed to real or simulated desalination effluents. It also describes toxicity tests performed within the context of environmental impact assessments (EIAs) of desalination plants (see Chapter 7). The studies were published in peer-reviewed journals and conference proceedings. Some studies belong to the gray literature, usually an EIA report prior to plant operations and local monitoring reports following plant operations. No review articles were included. Only the original publication was surveyed so that important details, often missing from review articles, could be compiled. Moreover, the laboratory bioassays and toxicity tests considered were limited to those associated with desalination. The vast literature on the effects of salinity and temperature on organisms, usually within the context of climate change, was not addressed.

This chapter is written with three objectives: (1) describe the actual impacts, if any, of desalination on the marine environment, as reported in the studies; (2) critically review the findings reported in those studies; and (3) generalize the findings, drawing attention to needed missing information. The chapter has four sections: (1) in situ studies, (2) in situ experiments and laboratory bioassays with natural local populations, (3) toxicity testing, and (4) a critical evaluation and integration of the findings.

Marine Impacts of Seawater Desalination: Science, Management, and Policy
https://doi.org/10.1016/B978-0-12-811953-2.00006-2

6.1 IN SITU STUDIES

In situ studies on the effects of seawater desalination on the marine environment were reported in 50 publications originating from 13 countries and 40 locations around the globe, spanning from 2001 to 2017 (Fig. 6.1). Table 6.1 summarizes the details on the area and date of the study, the specific desalination plant (start of operation, process, capacity, actual production), mode of brine disposal, stressor examined, and literature reference. These studies included abiotic and biotic impacts. As detailed in Chapter 4, abiotic impacts change the physicochemical properties or attributes of the receiving environment, including salinity, temperature, dissolved oxygen, particulate matter, turbidity, chemical composition, and sediment composition. The impacts are site-specific and highly variable. Biotic impacts change the living biota—abundance, physiological state, metabolic rates, and diversity—in all compartments of an ecosystem (see Chapter 3). Biotic effects may originate from a direct interaction with the brine or indirectly through environmental changes caused by abiotic impacts. Moreover, biotic effects depend on the specific tolerance and resilience of the organism. This complexity is presented in the studies described in this chapter, classified by geographical area and country of study. Some of the studies are described at length and some briefly, based on the availability of ancillary data. The details compiled in Table 6.1 complement the text.

6.1.1 The Gulf, the Red Sea, and the Gulf of Oman

More than half of the total global desalination effort takes place in the Gulf and Red Sea areas, which house, among others, the top two desalination countries: the Kingdom of Saudi Arabia (KSA) and the United Arab Emirates (UAE). Both the UAE and the KSA desalinate seawater from the Gulf, while KSA also desalinates seawater from the Red Sea. While thermal processes are the dominant technology in the area, recently, there has been an increase in installation and commissioning of plants using reverse osmosis (RO) technology. Contrary to vast number of publications on desalination technology from the area, only 12 studies reported actual in situ observations on the effects of brine discharge, with only three published before 2010 (Table 6.1). Three recent general hydrographic studies, incidentally mentioning desalination, are reviewed here as well.

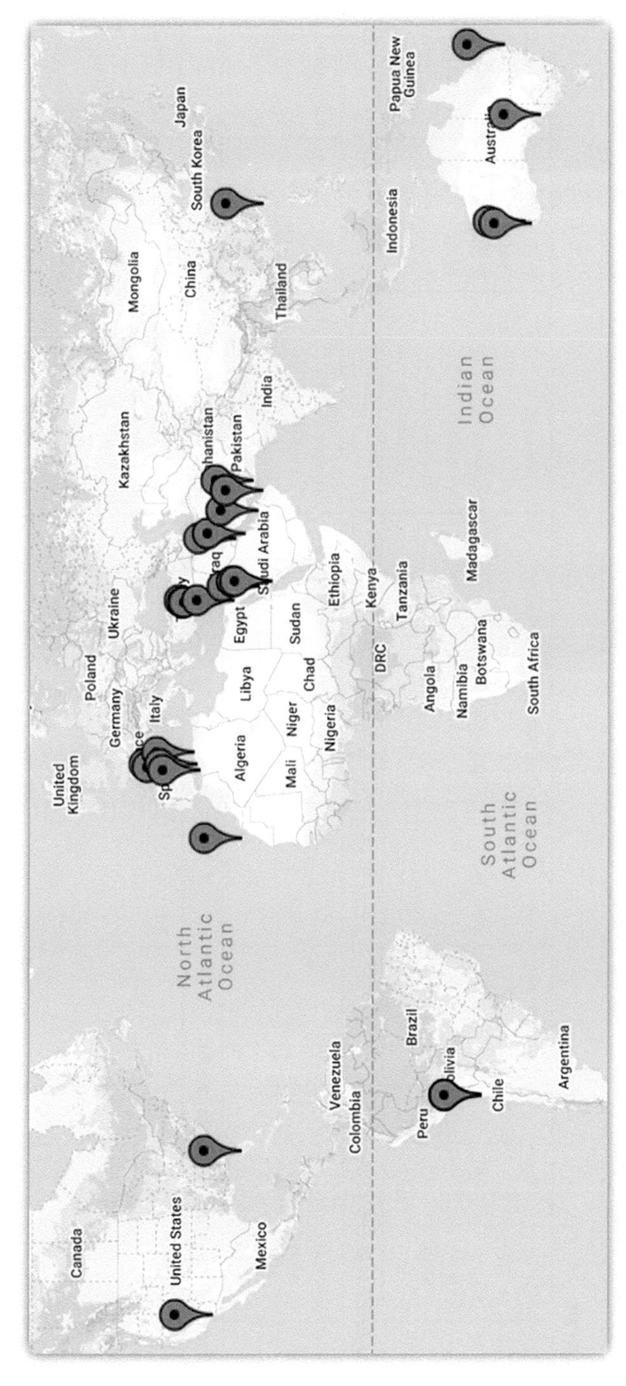

Fig. 6.1 Locations of studies (2001–17) reporting on observed impacts of seawater desalination on the marine environment.

Table 6.1 Compilation of details of in situ studies reporting on actual impacts of desalination on the marine environment

Location or plant name	Process	Start operations	Plant total capacity Mm^3/year (m^3/day)	Salinity brine, ambient	Discharge method (distance km; depth, m; D)	Year of study (actual production, % of total capacity)	Compartment (parameters measured)	Effect (distance, m)	Reference
The Gulf, Kuwait									
Az Zour	MSF	Before 1993	183 (501,000) in 2007	NR	Open MCWPP	1993–2009	Seawater (S, T)	Y (S, 5000) N (T)	Uddin et al. (2011)
Subiya	NR	NR	NR	NR, 36	Open MCWPP	NR	Seawater (S)	Y (S)	Uddin et al. (2011)
Kuwaiti Coast	General study					1992–2015	Seawater (T,S) Phytoplankton community	Y? (S) Y	Al-Said et al. (2017) and Al-Yamani et al. (2017)
The Gulf, United Arab Emirates (UAE)									
Abu Dhabi, Ras Al-Khaimah	General study					2013–14	Seawater (S, T, DO, pH, Tur) Phytoplakton community	Y? (S) N	Mezhoud et al. (2016)
The Gulf, Kingdom of Saudi Arabia (KSA)									
Ras Tanajib	SWRO	NR	NR	NR	Protected shallow coastal waters	NR	Sediment (metals)	Y (200)	Sadiq (2002)
Al-Jubail	MSF	NR	NR	NR	Open	1997–98	Phytoplankton and zooplankton communities	Y?	Abdul Azis et al. (2003)
Alkobar, Al Azizia	MSF	NR	NR	NR	ND	NR	Sediment (metals)	Y?	Alshahri (2017)
Al-Khafji	MSF	NR	NR	NR	ND	2017	Sediment (metals)	Y? (Cu)	Alharbi et al. (2017)

Al-Jubail	MSF, SWRO	NR	438 (1,200,000)	NR	Open	NR	Phytoplankton, Chl-a, nut, organic nut, TOC, metals, toxicity, bacterial growth and SPM	Y (phyto, organic nut, TOC, Cu)	Saeed et al. (2017)
							Metals in the emperor fish Lethrinus sp	N	
Red Sea, Kingdom of Saudi Arabia (KSA)									
Ash Shuqaya	MSF	1992	30.5 (83,400)	NR	Open	NR	Sediment (composition)	Y (7000)	Alharbi et al. (2012)
KAUST	SWRO	NR	14.6 (40,000)	80?, 40.9	Outfall (2.8, 18, N)	2012–13 (50%)	Seawater (S)	Y	van der Merwe et al. (2014a,b)
							Planktonic microbial population	N	
Marafiq Y1 complex	MSF, SWRO, MED	In stages	NR	NR	Open	2016	Seawater quality (S, T, Tur, SPM, Metals)	Y (S, T)	Ozair et al. (2017)
							Phytoplankton community	Y?	
							Various biotic studies (see text)	N	
Jeddah	MSF, SWRO	NR	0.16 (450)	NR	Open	NR	Phytoplankton, Chl-a, nut, metals, toxicity, bacterial growth, SPM	Y (phyto, organic nut, TOC)	Saeed et al. (2017)
Haql	SWRO	NR	1.6 (4500)	NR	NR	NR	Phytoplankton, Chl-a, nut, metals, toxicity, bacterial growth and SPM	Y (organic nut, TOC)	Saeed et al. (2017)

Continued

Table 6.1 Compilation of details of in situ studies reporting on actual impacts of desalination on the marine environment—cont'd

Location or plant name	Process	Start operations	Plant total capacity Mm^3/year (m^3/day)	Salinity brine, ambient	Discharge method (distance km; depth, m; D)	Year of study (actual production, % of total capacity)	Compartment (parameters measured)	Effect (distance, m)	Reference
Bay of Oman, Oman									
Al Ghubrah	MSF	1976	70 (191,160)	NR, 37.4–38.3 ($\Delta T = 5°C$)	Open, 3 discharges MCWPP	2004	Seawater (T, S, DO)	Y (T,S, DO,100)	Abdul-Wahab and Jupp (2009)
							Sediment (Metal)	Y (100)	
Barka	MSF	2003	33 (91,200)	NR, 37	4 pipes MCWPP (0.65, 2.5, N)	2004	Seawater (T, S, DO)	Y?	Abdul-Wahab and Jupp (2009)
							Sediment (Metal)	Y (100)	
Bay of Oman, Iran									
Chabahar Konarak	MSF	2006	5.5 (15,000)	46, 37.5	Open	2011	Seawater (S, T, pH)	Y (S, 300; T, pH)	Miri et al. (2015)
							Sediment (metal)	Y	
							Benthic Polychaeta	Y	Begher Nabavi et al. (2013)
Mediterranean Sea, Spain									
Platja de Mitjorn Formentera Island	GWRO	1985	0.73 (2000)	NR, 37.5	Outfall (0.02 0.9, N)	2001 (50%)	Seawater (S, pH, NOx, PO_4)	Y (50)	Gacia et al. (2007)
							Seagrass (*Posidonia oceanica*) and macrofauna	Y	
Blanes	SWRO	NR	10 (27,400)	60, NR	Outfall (NR, 19, D)	2002–03	Seawater (S)	Y (S, 10)	Raventos et al. (2006)
							Visual census of Macrobenthos	N	

Javea, Alicante	SWRO	2002	10.2 (28,000)	68, 37?	Open, preceded by dilution to S = 44 in a channel	2003–07 (50%, 25%)	Seawater (S) T?	Y (300)	Fernandez-Torquemada et al. (2009)
Alicante I	SWRO	2003	24.8 (68,000)	68, 37?	Open, preceded by dilution to S = 42–49	2004–08 (74–88%)	Seawater (S)	Y (2000) N if diluted	Fernandez-Torquemada et al. (2009)
				68, 37.9	Open	2004–05	Benthic communities	Y (500)	Del Pilar Ruso et al. (2007)
							Benthic Polychaeta	Y(500)	Del Pilar-Ruso et al. (2008)
Alicante II	SWRO	2009	24 (65,000)	NR, 37.7–38.3	Open, preceded by dilution with SW	2009	Seawater (S,T)	Y(750)	Garrote-Moreno et al. (2014)
New Channel of Cartagena, San Pedro del Pinatar, Murcia,	SWRO two plants	2005–06	24.8 (68,000) each	70, 37	Outfall, since 2006; with D since 2010 (5, 38)	2005–07 (various, see text)	Seawater (S)	Y (800- few km) N (after D)	Fernandez-Torquemada et al. (2009)
						2005–12	Benthic Polychaeta, S	Y (250) N (after D)	Del-Pilar-Ruso et al. (2015)
						2005–14	Benthic amphipoda, S	Y (250) N (after D)	de-la-Ossa-Carretero et al. (2016)
Mediterranean Sea, Algeria									
Mostaganem	SWRO	2011	73 (200,000)	68, 36.5	Outfall (1.4, 8, Y)	2014	Seawater (S) Benthic community	Y (200) Y	Belatoui et al. (2017)
Beni Saf	SWRO	2009	73 (200,000)	68, 36.5	Outfall (0.5, 8, N)	2014	Seawater (S) Benthic community	Y (1500) Y	Belatoui et al. (2017)
Bou Ismail	SWRO	2004	1.83 (5000)	NR, 38–39	NR	NR	Seawater (S, T, pH) Sediment (metals)	Y (S, pH)	Belkacem et al. (2016, 2017)

Continued

Table 6.1 Compilation of details of in situ studies reporting on actual impacts of desalination on the marine environment—cont'd

Location or plant name	Process	Start operations	Plant total capacity Mm^3/year (m^3/day)	Salinity brine, ambient	Discharge method (distance km; depth, m; D)	Year of study (actual production, % of total capacity)	Compartment (parameters measured)	Effect (distance, m)	Reference
Bousfer, Oran Bay	SWRO	ND	2 (5500)	64, 35.2	Open channel	2015	Brown algae (*Cystoseira compressa*) (metals)		Benaissa et al. (2017)
							Seawater quality (S, T)	Y	
							Limpet (*Patella rustica*) (Biomarkers)	Y (250)	
Mediterranean Sea, Israel									
Ashqelon	SWRO	2005	120 (329,000)	78, 39.5	Open MCWPP	2007–08	Seawater (S, T, DO, SPM, Tur, Fe, Nut)	Y (S, T, Fe, Tur, SPM)	Drami et al. (2011)
							Pelagic Bacterial community	Y	Drami et al. (2011)
						2012, 2014	Pelagic Bacterial community	Y	Belkin et al. (2017)
						2006–16	Seawater (S, T, DO, SPM, Tur, Nut)	Y (S, T, 3000) Y (TOP)	Shpir and Ben Yosef (2017b)
							Sediment (grain size, metal, C_{org})	N	
						2006–16	Benthic infauna	Y (600)	
Via Maris	SWRO	2007	30–90 (82,000–247,000)	62, 39	Outfall (0.6, 10, Y)	2005–13	Seawater (S, T, DO, SPM, Tur, Nut, TOC, Oil)	Y (S 1000) (T)	Kress and Galil (2012)
							Sediment (grain size, metal, C_{org})	N	
							Benthic infauna	Y?	
Via Maris	SWRO	2007	90 (247,000)	62, 39	Outfall (1.9, 20, Y), 800 m south of Soreq	2014–16	Seawater (S, T, DO, SPM, Tur, Nut, TOC, Oil)	Y (S, 3500; T, Tur, TOP)	Kress et al. (2016, 2017)
							Sediment (grain Size, metal, C_{org})	N	
							Benthic infauna	Y? (250)	

Soreq	SWRO	2013	150 (411,000)	62, 39	Outfall (1.9, 20, Y), 800 m north of Via Maris	2014–16	Seawater (S, T, DO, SPM, Tur, Nut, TOC, Oil)	Y (S, 3500; T, Tur, TOP)	Kress et al. (2016, 2017)
							Sediment (grain Size, metal, C_{org})	N	
							Benthic infauna	Y? (250)	
Hadera	SWRO	2010	127 (348,000)	NR, 39	Open MCWPP	2010–16	Seawater (S, T, O2, SPM, Turbidity, Nutrients	Y (S,T, TOP)	Shpir and Ben Yosef (2017a)
							Sediment (grain size, metal, Corg)	N	
							Benthic infauna	Y (600)	
Ashdod	SWRO	2016	100 (274,000)	NR, 39	Outfall (1.8, 22, Y)	2016	Seawater (S, T, O_2, SPM, Turbidity, Nutrients)	Y (S, TOP)	Shoham-Frider et al. (2017)
							Sediment (Grain Size, metal, Corg)	N	
							Benthic infauna	Y (0)	
Pacific/Indian Ocean, Australia									
Perth, Kwinana, Cockburn Sound	SWRO	2006	53 (143,700)	65, 33–37	Outfall (0.5,10,Y)	2006–12?	Seawater (S, T, DO)	Y (S, T 350)	Holloway (2009) and Bonnelye et al. (2017)
						2006, 2008, 2013	Benthic infauna and epifauna communities	N (3000)	Shute (2009) and Rivers (2013)
Gold Coast, Tugun	SWRO	2009	48.5 (133,000)	60, 37	Outfall (1.2, NR, Y)	2007–12 (57%, 28%, 7%) see text	Seawater (S, T, pH)	N	Viskovich et al. (2014)
							Sediment (grain size)	N	
							Benthic infauna and macroalgae	N? (60)	
Adelaide, Gulf of St. Vincent	SWRO	2011	100 (274,000)	NR, 35.9–37.4	Outfall (NR, 20, Y)	2012–14	Seawater (DO, pH and real time S and T)	Y (S, 100)	Kämpf and Clarke (2013) and Ayala et al. (2015)
							Macroalgae, benthic invertebrates, fish	N	
Southern, Binningup	SWRO	2012	100 (274,000)	59–64, 36	Outfall (0.75, 10, Y)	2015–16	Seawater (S, T, DO, pH, Tur)	Y? (S, T, 50)	Anon (2017)
						2016	Seagrass (*Posidonia augustifolia*)	Y? (300)	

Continued

Table 6.1 Compilation of details of in situ studies reporting on actual impacts of desalination on the marine environment—cont'd

Location or plant name	Process	Start operations	Plant total capacity Mm^3/year (m^3/day)	Salinity brine, ambient	Discharge method (distance km; depth, m; D)	Year of study (actual production, % of total capacity)	Compartment (parameters measured)	Effect (distance, m)	Reference
Pacific Ocean, Chile									
La Chimba, Antofagasta	SWRO	2003	18.9 (51,840)	56.7, 34.6	Outfall (0.24, 15, Y)	2002–12	Benthic communities (visual)	Y?	Vega and Artal (2013)
Pacific Ocean, Taiwan									
Penghu County	SWRO SWRO	2003, 2008	5.65 (15,500) trial	NR, 34	2 Outfalls (3, NR, NR)	2010–11	Seawater (S, T) Sediment (metal) Bivalve (metal) 50 m from outfall	Y (S) Y (Zn, Cu, 50) Y (Zn, Cu, 50)	Lin et al. (2013)
Atlantic Ocean, Canary Islands, Spain									
Las Burras, Gran Canaria	SWRO	1999	9.1 (25,000)	50?, 36.7	Outfall (0.3, 7, NR)	2008–09	Seawater (S) Sediment (grain size and OM) Benthic macrofauna and meiofauna	Y (S, 30) N Y	Riera et al. (2011, 2012)
Maspalomas II, Gran Canaria	SWRO	1988	9.3 (25,000)	69.5, 36.8	Outfall (0.3, 4), with D since 2011	Before and after 2011	Seawater (S, DO, pH, sodium metabisulfite (SMBS)) Soft bottom fish	Y (S, 700) Y (pH, DO, with SMBS) Y	Portillo et al. (2014)
Atlantic Ocean (Gulf of Mexico), United States									
Tampa Bay	SWRO	2003	35 (95,900)	54–62, 21–30	Open, MCWPP	2003–07? 2002–09	Seawater (S, T) Benthic macro invertebrate	Y (S,T) N	McConnell (2009) and PBSJ (2010)

?, inconclusive; *Chl-a*, chlorophyll-a; C_{org}, organic carbon; *D*, diffuser; *DO*, dissolved oxygen; *MCWPP*, mixed with cooling waters from power plant; *N*, no; NO_x, inorganic nitrogen; *NR*, not reported; *nut*, nutrient; *OM*, organic matter; *phyt*, phytoplankton; PO_4, phosphate; *S*, salinity; *SMBS*, sodium metabisulfite; *SPM*, suspended particulate matter; *T*, temperature; *Tur*, turbidity; *Y*, yes.

6.1.1.1 Kuwait

The Az Zour multistage flash distillation (MSF) desalination and power plant started to operate before 1993 and was expanded in 2005. In 2007, the plant had a total capacity of 501,400 m^3/day. Brine was co-discharged with the cooling waters of the power plant through an open outfall. From 1993 to 2003, surficial seawater salinity increased from 36–41.5 to 41–43 with no concurrent change in temperature. Salinity further increased in 2007–09, with peaks salinities of 50 ca. 5 km from the discharge area, partly attributed to the expansion of the plant. The natural seasonal variation in salinity, significant until 1996, had a much lower variability in the later years of the study. Salinity was also high at the discharge channel of the Subiya desalination and power plants with a value of 40 compared to the ambient 36. The low ambient salinity was due to the freshwater discharge from the Shatt Al-Arab river, while the higher salinity at the discharge site was attributed to desalination since there were no other coastal activities in the area.

Long-term measurements of sea surface salinity and temperature were collected from 1982 to 1990 and from 1994 to 2015 in Kuwait Bay (8.5 m depth) and in open waters (26.5 m depth, 40 km offshore). Salinity in the bay increased from 37.6 to 44.9 and at the offshore station from 36.4 to 43.1 between 2006 and 2011. This increase was attributed to reduced river flow from the Shatt Al-Arab, to increased evaporation, and to brine discharge from four desalination and power plants (Subiya, Doha, Shuwain, and Shuaiba). The relative contribution of brine discharge to the increase in salinity was not addressed. No significant shifts were observed in temperature, having an annual long-term mean of 23.6°C and a marked seasonal variability (9.7–33.7°C). Hypersalinity substantially shifted the phytoplankton's composition, abundance, and community structure. The number of taxa in Kuwait Bay decreased from 353 in 2000–07 (69% diatoms and 29% dinoflagellates) to 159 taxa (58% diatoms and 36% dinoflagellates) in 2012–13. At the open sea station the number of taxa was similar, but the relative distribution differed between the periods: 164 (66% diatoms and 29% dinoflagellates) in 2005–06 and 156 (53% diatoms and 43% dinoflagellates) in 2012–13. Small-sized phytoplankton species contributed significantly to the community at both stations, with higher numbers in the latter years. Chlorophyll-a (Chl-a) concentrations decreased by a factor of 2, from about 6 μg/L in 2000 to 3 μg/L in 2009 and remained low until 2013. A drastic decrease in fish catch landing was observed from 2000 to 2012 as the phytoplankton decreased, illustrated by the fish species *Tenualosa ilisha* that feeds on plankton.

6.1.1.2 United Arab Emirates

A general study, not addressing a specific desalination plant, was conducted in 2013–14 off Abu Dhabi and Ras Al Khaimah. Salinity, temperature, dissolved oxygen (DO), pH, secchi depth, Chl-a, and phytoplankton were measured along transects from the shoreline to the offshore. Salinity decreased from the onshore toward the offshore, probably due to the stronger evaporation and lower precipitation at the coastal areas. However, contribution of brine discharge from the nine main desalination plants along the coast was also considered to increase salinities close to shore. Surficial temperature did not exhibit a clear spatial gradient; however, Chl-a, DO, pH, and secchi depth increased with distance from the shore. No spatial trend or significant dominance of specific species in the phytoplankton communities were identified, except for the dominance of diatoms off Ras Al Khaimah.

6.1.1.3 Kingdom of Saudi Arabia

Phytoplankton and zooplankton were studied monthly at the vicinity of the Al Jubail MSF plants from 1997 to 1998. Primary production was higher in the coastal area receiving the discharges from the outfall bay. While the phytoplankton species composition did not change significantly among the sites, the abundances of diatoms, dinoflagellates, and cyanobacteria were lower at the outfall bay in comparison to the open sea and the intake. Seasonality was found as the main factor affecting biomass.

The seawater RO (SWRO) plant at the King Abdullah University of Science and Technology (KAUST) in the Red Sea had an installed capacity of 40,000 m^3/day. Brine (S=60) was discharged through an outfall equipped with 4 discharge screens, ca. 2.8 km from the shoreline, at 18 m depth. At the time of the study (2012–13) the plant was operating at half capacity. Salinity at the outfall was ca. 47 compared to ambient 40.9, measured at stations 25 m away. Planktonic microbial abundance was lower at the more saline stations and increased with increased distance from the discharge and decrease in salinity. As the bacterial abundance in the brine was lower than ambient, this finding was attributed to a dilution effect rather than a direct impact of the brine discharge.

The Marafiq-Y1 power and desalination plants (MSF, multieffect distillation (MED), and SWRO) are located within a large complex spanning 20 km along the KSA Red Sea coast. The plants started operation in several stages with the oldest in operation for >30 years. In 2016, the average annual concentrations in the brine were: residual Cl_2 0.0372 mg/L, Cu 25.6 μg/L, excess salinity over ambient by 1.14, temperature 35.53°C and pH 8.03.

Very limited impact of the brine was recorded. Slightly elevated parameters rapidly returned to ambient values due to cascade mixing at the brine discharge channel and to hydrographical conditions. Metals and organic compounds in fish tissues were similar in specimens from the discharge channel and from a reference site. Dead and live corals were found in all areas, indicating that the brine had no acute impact nor posed a long-term threat to marine life. Brine reduced phytoplankton density and primary production within a limited area next to the discharge, with rapid recovery away from it. No effect was found in the overall species composition. The elevated temperature and salinity did not affect the occurrence of zooplankton.

In a study published in 2007, biotic effects were studied at the brine discharge sites of three desalination plants: Al-Jubail (MSF and SWRO, open discharge) in the Gulf and Jeddah (thermal and SWRO, open discharge) and Haql (SWRO, outfall) in the Red Sea. Temperature and salinity data were not provided. Plankton, chlorophyll production, nutrients, trace metals, toxicity, bacterial growth, and total suspended solids, measured at the discharge site of each plant were compared to the values found at the respective intake site, representing reference conditions. The results showed essentially no impact of brine discharge in most cases and, when identified, the impact was minor. Phytoplankton numbers and Chl-a were higher at the discharge site in the Jeddah and Jubail plants compared to the intake but similar at both sites in Haql. Nutrient concentrations were similar at both sampling areas in Jeddah. In Jubail, nutrient concentrations at the discharge site were lower, probably a result of utilization by diatoms and macrophytes that grew in the discharge channel. Bacteria numbers were similar at the intake and discharge sites of Jubail and Jeddah. Organic nutrients and TOC were higher at the discharge sites compared to the respective intakes, a result of organic matter decomposition due to chlorination. In all plants, metals concentrations at the intake and discharge sites were similar. Only copper in the discharge channel at Jubail was higher than in both the intake and discharge sites, presumably originating from corrosion of the MSF plants. No metal accumulation was detected in the emperor fish (*Lethrinus* sp.) collected from both sites in Jubail.

The composition and metal content of marine sediments along the Gulf and Red Sea coasts of the KSA were determined at the discharge sites of four desalination plants. In the Gulf, metals concentration decreased with increased distance from the outfall of the Ras Tanajib SWRO plant, with the impact area confined to 200 m. Elevated metal concentrations were also found at the discharge site of the Alkhobar MSF plant, while the metal concentrations off the Khafji MSF plant were natural for the area except for copper, which was higher near the discharge site. High chlorite levels were measured in marine sediments at the vicinity of the Ash Shuqaya MSF plant in the Red Sea, up to 7 km north and south of the plant and 5 km seawards.

6.1.1.4 Sultanate of Oman

The Al Ghubrah (Muscat) MSF and power plants started to operate in 1976 and had a total capacity of 191,200 m^3/day. Brine and cooling waters were discharged at the shoreline at three locations. The Barka MSF and power plants, located 65 km NW of Muscat, started to operate in 2003 and had a total capacity of 91,200 m^3/day. Brine and cooling waters were discharged through 4 pipelines, 0.65 km from the shoreline at 2.5 m depth. During the study conducted in 2004, there was only a slight increase in temperature and salinity near the outfall of the Barka plant while near the Al Ghubrah discharge site, temperature and salinity were higher than ambient (35°C compared to ambient 30–31°C, and salinity of 40 compared to ambient 37.4–38.3). Dissolved oxygen was lower at the affected stations due to lower solubility. Copper concentrations in the sediments and to a lesser degree, lead, cadmium, and zinc, were higher at stations near the outfalls and attributed to the brine discharge. The effects were detected at the outfall and up to 50–100 m distance.

6.1.1.5 Islamic Republic of Iran

The Chabahar Konarak MSF plant started to operate in 2006 and discharged the brine (S = 46) through an open outfall at the shoreline. At the time of the study in 2011, the plant's total capacity was 15,000 m^3/day. Sediment samples for the determination of metals and Polychaeta community structure were collected prior to and after the winter monsoon season. The farthest station was located 1.2 km from the discharge channel. Average salinity at the discharge site was 45.5 and decreased with increased distance from the discharge, remaining higher than ambient (S = 37.5) up to 300 m from the discharge. Temperature was higher than ambient at the discharge site by 3°C and 1°C in the pre- and postmonsoon survey, respectively. At the discharge site pH was low (7.15). The sediments were homogenous and dominated by silt and clay fraction. Metal concentrations in the sediments generally decreased with increased distance from the discharge area, but the highest concentrations were measured not at the discharge site but 200 m away. Lower pH and higher salinity and turbulence at the discharge site probably contributed to the solubility and dispersion of metals introduced with the brine, preventing accumulation in the sediment. Only

Cu was higher at the discharge station. Polychaeta abundance was highest at the reference station and lowest at the discharge area during both surveys, and community structure was influenced by temperature and salinity. The most frequent family found was the Spionidae showing the tolerance of this family to brine discharge.

6.1.2 Mediterranean Sea

The second largest area in seawater desalination effort is the Mediterranean Sea. Spain is the fourth largest producer of desalinated water, while desalination in Algeria, Israel, Libya, Tunisia, Malta, and Cyprus is substantial compared to the natural freshwater availability in these countries. RO is the most common desalination process in the area (>80%).

6.1.2.1 Spain

The Javea SWRO plant started to operate in 2002 using coastal wells for feedwater supply. The plant had a total capacity of 28,000 m^3/day but during the time of the study (2003–07) operated at half capacity in the summer and at a quarter capacity in the other seasons. Brine (S = 68) was mixed with seawater (S = 37) to a salinity of 44 prior to its discharge through diffusers into the artificial La Fontana Channel and subsequently to the sea. The water salinity at the Fontana channel never exceeded 44. At the shoreline, brine dispersed near the bottom. Maximal salinity during the summer was 42–43, and the elevated salinities could be traced up to 300 m from the discharge site. During the rest of the year, the effect of the brine discharge on salinity was localized in the channel.

Three studies were conducted at the Alicante I SWRO plant and one after the addition of Alicante II SWRO plant. Alicante I started to operate in 2003 and used beach wells for feedwater supply. The initial total plant capacity of 50,000 m^3/day was increased in 2006 to 68,000 m^3/day but actual production was 60,000 m^3/day. Brine (S = 68) was diluted with 1.5–5 parts seawater and discharged at the shoreline. In 2009, Alicante II started to operate in the area with a total capacity of 65,000 m^3/day, discharging brine at the shoreline after dilution with seawater. During 2004–08, only a small increase in salinity was detected at the surface waters; however, salinity near the bottom was higher than the ambient (40 and 37, respectively). Increased salinity was observed up to 2 km from the discharge point. In the summer, the maximum salinity was found at the seasonal thermocline at 13 m. When the brine was diluted with seawater to S = 43 prior to discharge, there was almost no salinity signal in the receiving environment.

The benthic community assemblages, emphasizing the Polychaeta families, were studied in 2004–05 at three stations located at 4, 10, and 15 m depth opposite the discharge site. Salinity at 4 and 10 m depth was 40.6 and 38.9, respectively, higher than ambient. Abundance and diversity near the discharge were lower than at the more distant stations. Five months after Alicante I started operations, the most abundant taxa at the station near the discharge were the Polychaeta (65%), followed by Mollusca (20%) and Crustacea (15%). After 9 months of operations, the nematodes comprised 50% of the infauna, and after 21 months, nematodes dominated the community at stations with salinity higher than 39, reaching up to 98% of the total community. Some of the Polychaeta families, such as Paraonidae, tolerated higher salinities while others decreased in abundance and even disappeared near the discharge area. Some of the changes in communities were also attributed to differences in water depth and sediment composition. Following the operation of Alicante II, the salinity at ca. 750 m from the discharge site, at 10 m depth was higher by 1–2 than ambient and temperature higher by 0.5–1°C. Salinity at the affected stations fluctuated in a few hours over a range of 2.

The two SWRO plants at the New Channel of Cartagena started to operate in 2005 and 2006 with a total capacity of 68,000 m^3/day each. One plant used horizontal wells and one an open intake for feedwater supply. Since January 2006, brine (S = 70) was discharged through a 5 km long outfall at 38 m depth. The outfall was equipped with a diffuser only in 2010. From May 2005 to January 2006, prior to the completion of the outfall, brine (7000–28,000 m^3/day) was diluted with seawater and discharged at 2 m depth. While the brine was discharged at 2 m, the maximal salinity measured was 38.2 (compared to ambient 37), with brine dispersing near the bottom up to 800 m from the discharge point. The dispersion pattern when brine was discharged through the outfall without the diffuser varied with season and volume of brine. Although not clear from the study, it seems that salinity reached 44 at the outfall near the bottom. At times this increase was localized and at least in one realization, the salinity signal was found a few kilometers from the outfall. Following the installation of the diffuser system, salinity near the outfall was similar to the ambient. Benthic assemblages of Polychaeta and Amphipoda were studied in the outfall area: in 2005, prior to the brine discharge; from 2006 to 2009, with brine discharged through the outfall without a diffuser; and since 2010, with the diffuser. The bottom salinities at the outfall prior to brine discharge and after the installation of the diffuser were the ambient salinity (37). Without the diffuser, the salinities were higher than ambient and increased with the increase in discharge from

39.5 to 48.8 at the outfall and from 38.5 to 43, 250 m east of the outfall. During hypersalinity, the Polychaeta and Amphipoda abundance and diversity were lower at the outfall and to a lesser degree at the station located 250 m eastwards. Paraonidae and Magelonidae were tolerant to hypersalinity and found at the outfall. The abundance of other families decreased and later recovered, at different rates, following the introduction of the diffuser and the return to ambient salinities. Similarly, the presence of the amphipods *A. diadema*, *A. typica*, or *P. longipes* at the station closest to the outfall during the higher salinity period indicated tolerance to hypersalinity. Amphipod abundance increased 6 months after the diffuser installation, with rates of recovery dependent on the species.

Meadows of the seagrass *Posidonia oceanica* were present at the vicinity of the outfall of the Platja de Mitjorn GWRO plant, where brine dispersed close to the bottom, measurable up to 50 m from the outfall. A study was conducted in 2001 at an impact site ($S = 37.8$–39.8) and at a far field impact site ($S = 37.8$–39.3). Ambient salinity was 37.5. At the outfall, salinity (41.5) and pore water salinity ($S = 38.0$–40.1) were higher than ambient values, pH (7.5) was lower and NO_x and PO_4 concentrations (20.8 and 0.54 μM, respectively) much higher than ambient. The *P. oceanica* meadows at the impact site revealed low shoot abundance, significant reduction in leaf size, and overload of epiphytes (Fig. 6.2). However, the main cause of degradation in this area was assumed to be eutrophication due to high nutrient concentrations, high epiphyte abundance, and high herbivore activity. At the far field impact site, no differences were found in shoot densities, but there were changes in the structural pattern of the shoot distribution, increase in the frequency of necrosis marks in the leaves, and a significant lower abundance of the accompanying macrofauna (echinoderms, holothurians, and sea urchins). Since no eutrophication symptoms were detected at the far impact site, the effects were attributed to salinity stress. In general, there was no indication of extensive decline of the affected meadow.

Visual censuses of the macrobenthos (fish, echinodermata, mollusca, polychaeta, and decapoda crustacea) were conducted by scuba divers at the brine outfall of the Blanes SWRO plant prior to (2002–03) and following (2003–04) plant operations. Salinity higher than ambient was detected up to 10 m from the diffuser. No significant variations attributable to the brine discharge were detected in the abundance, species richness, and diversity of the macrobenthos, probably due to the rapid dilution of the brine and the high natural habitat variability. The frequent sightings of fishes above the outfall indicated that the pipe acted as an attractor for fish, similar to artificial reefs.

Fig. 6.2 *Posidonia oceanica* meadow (A) protected by the EU Habitat Directive and Barcelona Convention (see Chapter 7). (B) Schematic representation of the plant and (C) its fruits and seeds. *Reproduced from (A) https://commons.wikimedia.org/w/index.php?curid=2673280; and (C) https://commons.wikimedia.org/w/index.php?curid=10571505.)*

6.1.2.2 Algeria

The Bou Ismail SWRO desalination plant started to operate in 2004 and had a total capacity of 5000 m^3/day. At the time the papers were written (2016) the plant was not operational. Brine discharge did not affect water temperature at the discharge site. Salinity was ca. 43, higher than the ambient 38–39, and seawater pH was low (6.4) due to the acidic discharge and high phosphate concentration from the polyphosphate antiscalant. The brine discharge did not affect the metal concentrations in the sediments nor in the brown algae *Cystoseira compressa*.

The Beni Saf and Mostaganem SWRO plants started to operate in 2009 and 2011, respectively, each with a total capacity to produce 200,000 m^3/day fresh water. Beni Saf's brine was discharged through a single port outfall at 8 m depth, 500 m from the shoreline, while Mostaganem's brine was discharged through an outfall equipped with diffusers, at 8 m depth, 1400 m from the shoreline. The increase in salinity and the dispersion of the brine measured in 2014 were different at the two outfalls; the maximal salinities were 62.8 and 39.8 and Beni Saf and Mostaganem, respectively, compared to the ambient salinity of 36.5. Salinity over 38 was detected >1.5 km from the discharge point at Beni Saf and only 200 m at Mostaganem. At both sites, brine reduced the abundance and diversity of the benthic communities near the outfalls. The impact was higher at Beni Saf due to lower dilution, with most species disappearing near the outfall. Only the Polychaeta from the families Spionidae and Paraonidae, the amphipoda (crustacean) species *Urothoe grimaldi*, *Synchelidium haplocheles*, *Periculodes longimanus*, the mollusk *Chamelea galina*, and the group of ribbon worm Nemertea were capable of surviving near the outfall but in very low abundance.

The Bousfer SWRO plant, with a total capacity of 5500 m^3/day operated mainly in the summer and discharged the brine through an open channel. In 2015, salinity at the outfall (43.5) was higher than at stations 150–250 m away, which exhibited ambient salinity (S = 35.2). Temperature at the outfall was lower than ambient (23.4°C and 25.6°C, respectively). Possible neurotoxicity, genotoxicity, oxidative stress, and damage of the brine were checked in limpets (*Patella rustica*). All antioxidant defense enzyme activities, except for one, as well as neurotoxicity and genotoxicity markers, were significantly higher in the limpets collected from the brine discharge area indicating oxidative stress. Oxidative damage was also higher but not significantly different. Molecular damages were observed in limpets from areas adjacent to the discharge site, probably due to chronic exposure to the brine.

6.1.2.3 Israel

The Ashqelon SWRO plant has been operating since 2005 with a total plant capacity of 329,000 m^3/day. Feedwater (S = 39.5) is supplied through a submerged intake, and brine is discharged at the shoreline next to the cooling waters from the Rutenberg power plant at a maximal volume ratio of 1:25 (brine:cooling waters). Until 2010, the backwash of the sand prefilters, containing Fe salts used as a coagulant in the pretreatment stage, was discharged in pulses with the brine. Since 2010, the backwash is discharged continuously with the brine. Additional co-discharges include brine from well amelioration RO plants and cooling waters from a gas-powered power station. Long term annual monitoring, conducted 2–4 times a year has detected increases in salinity up to 4 and of temperatures up to 8°C near the discharge area. The extent and direction of the mixed brine-cooling waters dispersal was highly dependent on the season, hydrography and on the operation of the plants during sampling, reaching up to 3 km from the outfall, mostly near the surface but also near the bottom. The discharges did not affect dissolved oxygen saturation (DO_{sat}), Chl-a concentration nor nutrient concentrations, except for total organic phosphorus (TOP), which was higher at the more saline stations. The elevated TOP originated from the polyphosphonate based antiscalant used in the process and discharged with the brine. Benthic community structure was naturally affected by the season, bottom depth, and sediment grain size. However, stations near the discharge had lower abundance and different biotic community structure than the more distant stations, indicating an effect of the discharges up to 600 m from the discharge site. Dedicated in situ research conducted in 2007–08, 2012, and 2014 found similar increases in salinity and temperature as the monitoring studies. However, although seawater turbidity since 2010 has been mostly unaffected by the discharges, when the backwash was discharged in pulses, turbidity increased rapidly, from 0.5 to 4 NTU, increasing the concentrations of particulate matter and iron in seawater. A conspicuous "red plume" was formed, creating a unique visualization of its aesthetical impact (Fig. 6.3). Moreover, elevated Fe in seawater reduced the photosynthetic efficiency of autotrophs. Chl-a concentrations, pelagic microbial

Fig. 6.3 A unique visualization of the marine environmental impact of effluents discharge from a seawater desalination plant, Ashqelon, Israel. Until 2010, the backwash of the sand prefilters, containing Fe salts used as a coagulant in the pretreatment stage, was discharged in pulses with the brine and tinted the seawater red. Since 2010, the backwash is discharged continuously with the brine, a mitigation measure that prevents this occurrence. (A) Open brine discharge channel (marked by the *arrow*), next to the power plant cooling water discharge; (B) areal dispersion of the "red plume," with the power plants' coal unloading quay on the background; (C) natural seawater color in the area; (D) turbid red seawater following the discharge of the backwash. *Courtesy of Yaron Gertner, IOLR.*

abundance, and primary production were lower and bacterial production and cell specific bacterial activity higher at the affected stations. The magnitude of the effect was seasonally dependent, with higher impact in the summer. Microbial community composition changed as well. In the summer, the diatoms comprised the principal group of the eukaryotic community at high salinities while in the winter there was a lower abundance of cyanobacteria and dinoflagellates at high salinities.

The Palmachim and Soreq SWRO plants have been operating since 2007 and 2013, respectively. Both plants used a submerged intake system for feedwater supply. The Palmachim plant started operations with a total installed capacity of 82,000 m^3/day, discharging the brine (S = 62, ambient 39) through a marine outfall equipped with diffusers 0.6 km from shore, at 10 m depth. The capacity of the plant was expanded gradually to 246,600 m^3/day in 2013. Since April 2014, the brine has been discharged through a marine outfall equipped with diffusers 1.4 km from the shoreline, at 20 m bottom depth. The Soreq plant has an installed capacity of 411,000 m^3/day. Brine (S = 62, ambient 39) is discharged through a marine outfall equipped with diffusers 1.9 km from the shoreline, at 20 m depth. The distance between the outfalls of the two plants is 0.9 km. Background surveys were performed at the EIA phase, and biannual monitoring studies have been performed since the start of operations. From 2007 to 2013, while brine from the Palmachim plant was discharged at 10 m depth, high salinity was measured at the outfall, near the bottom (up to 45, ca. 15% above background). The hypersaline seawater was confined to the bottom 1–2 m water layer and detected even at a distance of 1 km from the outfall, encompassing an area of approximately 0.4 km^2. Seawater was slightly warmer (ca 0.5°C) at the outfall, near the bottom. No effect of the brine was detected on DO_{sat}, in particulate matter, nor in Chl-a concentrations. Metals concentrations in sediments were natural for the studied area and not affected by the discharge. Benthic infauna assemblages varied temporally and seasonally. Higher Polychaete and opportunistic taxa were found near the outfall during some of the surveys. The temporal data were insufficient to establish a definite effect, if any. Monitoring at 10 m depth was discontinued following the cessation of brine discharge through this outfall. Monitoring at the two 20 m depth outfalls in 2015–16 found salinities up to 41.76 (ca. 7.5% above background). The brine was negatively buoyant, mainly confined to waters 1–2 m above the bottom but observed also at 4.5 m above it. The brine signature was found up to 3.5 km from the outfalls. The hypersaline layer (salinity at 1% above reference) encompassed an area up to 15 km^2, highly dependent on the hydrographical conditions and discharge rate during the survey. The brine increased seawater temperature by ca. 0.5°C as well as the turbidity and did not affect the concentrations of DO_{sat}, Chl-a, total organic carbon, nor nutrients in seawater. Only TOP concentration was higher than ambient and linearly correlated to salinity (Fig. 6.4). TOP originated from the polyphosphonate-based antiscalants used in both plants and discharged with the brine at sea. Metals and organic carbon concentrations in the sediments were low and natural for the area at all stations. The benthic infauna assemblages (number of individuals, number of taxa, diversity index, and species evenness) studied at stations located at the outfall and up to 2200 m distance varied seasonally and temporally. The biotic findings showed no brine impact on the infauna at stations located at 200 m or more from the outfalls. At the two outfall stations and a few stations located within 200 m of the outfalls there were indications of impact but with conflicting results. Some of the statistical analyses indicate an effect of the brine on the biota while others did not.

Two additional SWRO plants operate along the Mediterranean Coast of Israel: Hadera, since 2010, and Ashdod, since 2016. Hadera's total

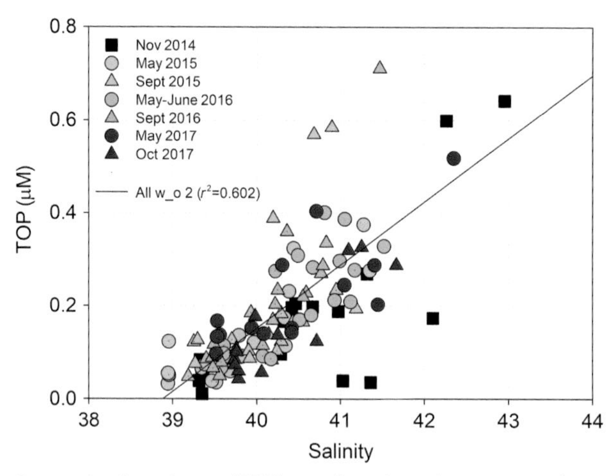

Fig. 6.4 Total organic phosphorus (TOP) as a function of seawater salinity at the brine discharge area of the Palmachim and Soreq (Israel) desalination plants during seven operational monitoring surveys. Organic phosphorus originates from the polyphosphonate-based antiscalant used in both plants and discharged with the brine. *(From Kress, N., Shoham-Frider, E., Lubinevski, H., 2017. Monitoring the Effect of Brine Discharge on the Marine Environment: A Case Study off Israel's Mediterranean Coast. In: The International Desalination Association World Congress on Desalination and Water Reuse 2017, Sao Paulo, Brazil (open source, in Hebrew).)*

production capacity is 348,000 m^3/day and discharges the brine at the shoreline after dilution with cooling waters from the Orot Rabin power plant. Ashdod's total production capacity is 274,000 m^3/day and discharges the brine through a marine outfall equipped with diffusers at 1.8 km from the shoreline at 22 m depth. Biannual marine environmental surveys are conducted at both discharge areas. Six years of monitoring at Hadera's discharge area have shown no effect of the brine-cooling water in most parameters measured (turbidity, DO_{sat}, Chl-a, nutrients, organic carbon and metals in the sediments), except for TOP, which was higher at the more saline stations. Seawater temperature was up to 5°C and salinity up to 6% above ambient levels. The warm and hypersaline discharge plume dispersed either at the surface or near the bottom to variable distances (up to 3 km) and directions from the discharge site, depending on the number of units working at the plants. TOP originated from the polyphosphonate based antiscalant used in the process and discharged with the brine. Benthic community structure within 600 m of the discharge site had lower abundance and different structure compared to the reference stations. One year of monitoring at the Ashdod's SWRO brine outfall showed that the brine dispersed near the bottom. Salinities up to 2.5% higher than ambient were found up to 500 m from the outfall. TOP, originating from the polyphosphonate antiscalant, was linearly correlated to salinity. Benthic community structure was different and abundance lower at the outfall compared to the reference stations.

6.1.3 Pacific and Indian oceans

6.1.3.1 Australia

The Perth (Kwinana) SWRO plant has been operating since November 2006 with a total capacity of 143,700 m^3/day. Brine ($S=65$, ambient 33–37) is discharged through an outfall equipped with a diffuser 500 m offshore at 10 m depth. Bottom salinity near the outfall (ca. 350 m westwards) increased by 1 (3% over reference), and a slight temperature stratification was observed. Brine did not impact DO_{sat} in the area. The benthic infauna and epifauna communities were characterized prior to (2006) and following (2008, 2013) plant operations. Multiple stations were sampled at each of four different zones: the potential impact zone (>3 km east of the outfall) and three reference zones (northern, central-west, southern). During all three surveys, infauna communities were dominated by Polychaeta and Mollusca. Regional observed changes in community structure were strongly correlated with bivalve abundance that were most abundant during the baseline survey and absent in 2013. The number of taxa, richness, and diversity at stations in the potential impact zone during each survey were similar to the reference zones, indicating no impact of brine discharge. Regional changes in community structure of the epibenthos among surveys were correlated to the abundance of tube anemones and horseshoe worms (present largely in the baseline and 2008 surveys) and sponges (most abundant in the 2013 survey). Epibenthic taxa were highly variable among sites in 2013, primarily due to a large abundance of sea pens at the southern reference zone and sea cucumbers at the northern reference zone. Changes in infauna and epibenthic communities in 2013 were linked to the regional increases in phaeophytin and Chl-a, while differences in community structure between the baseline and 2008 surveys were linked to regional declines in coarse sand and increases in silt and clay. No effect was attributed to the brine discharge.

The southern SWRO plant, Binningup, operated from 2012 to 2013 with a total capacity of 137,000 m^3/day and has been operating since January 2014 with a capacity of 274,000 m^3/day. Brine (S = 50–65, ambient 35–37) is discharged through an outfall equipped with diffusers, 750 m from the shoreline at 10 m bottom depth. During 2015–16, the bottom salinity at 50 m from the outfall ranged from 35.5 to 37.5. Temperatures were slightly higher near the outfall and DO_{sat} was never lower than 60%, the trigger value for action. Turbidity and pH were natural for the region. Benthic health monitoring surveys of the seagrass *Posidonia angustifolia* found a decrease in shoot density from 2012 to 2016 at two of the four stations at the impact site (300 m west of the diffuser) as well as at two stations at the southern reference site. Median shoot density at the impact site was lower than the environmental quality standard when compared to the reference stations at 5 km from the outfall but not when compared to the reference stations at 1.8 km from the outfall.

The Gold Coast SWRO started to operate in 2009 with a total capacity of 133,000 m^3/day. The brine (S = 60, ambient 37) was discharged through an outfall equipped with diffusers at 1.2 km offshore. During the time of the study (2007–12), the actual production of the plant decreased from 57% (2009–10), through 28% (2010–11), and down to 7% (2011–12). Sediment samples for the determination of grain size, taxonomic richness, and abundance of benthic infauna and macroalgae assemblages were collected biannually at 60 m from the outfall and at two reference sites located 500 m away. The composition and particle size distribution of the sandy sediment remained similar during the study at the three sites. Three years after limited operation, the benthic assemblages at the impact location differed from those at the northern and southern reference locations, which did not differ.

However, the magnitude of the difference among locations was much smaller than the differences among sampling periods for each location. Since salinity, temperature, and pH were similar, the differences between impact and control locations found in 2012 were attributed to natural variation rather than to an impact of the desalination brine.

The Adelaide SWRO plant became operational in late 2011, producing 137,000 m^3/day and at full capacity (274,000 m^3/day) in December 2012. The brine was discharged through a 1.1 km-long tunnel terminating with 6 risers equipped with diffusers at 20 m depth. Four CTDs instruments, located at a radius of 100 m from the outfall and a fifth at the intake measured salinity and temperature and transmitted the data in real time to the plant. Ambient salinity from 2012 to 2014 was 35.9–37.4. Maximal salinities, measured 100 m from the outfall, were higher than ambient by 0.6–1.1. Brine discharge did not affect DO concentrations or seawater pH. Macroalgae coverage, benthic invertebrates, mobile invertebrates, and fish were counted and identified by scuba divers. Assemblages of invertebrates and algae were similar at the discharge and reference areas. Diversity and abundances of fish differed among sites and seasons but not among sites within one season. The variability observed could not be attributed to the construction and/or operation of the desalination plant.

6.1.3.2 Chile

The La Chimba SWRO plant started to operate in 2003 with a total production of: 12,930 m^3/day in 2005; 25,920 m^3/day in 2008; and 38,880 m^3/day in 2010. Since 2011, it has been producing 51,840 m^3/day. Brine ($S = 56.7$, ambient 34.6) was discharged through an outfall equipped with a diffuser at a depth of 15 m. The study took place prior to (2002) and following (2005, 2008, 2010, 2012) plant operations. No data on in situ salinity were provided. Benthic macrofauna communities (coverage, identification, counts) were determined by divers along five transects parallel to the shoreline at 0–100 m from the outfall. Individual organisms were collected for further identification and biomass determination. In 2010, a bank of the ribbed mussel *Aulacomya atra* was detected for the first time at the brine discharge area. Adult specimens appeared closest to the discharge, and younger specimens appeared more distant from the discharge. It was speculated that the higher salinity around the outfall created conditions appropriate for the settlement of this species that were absent prior to the operation of the plant. A proliferation of predatory species such as echinoderms (especially starfish) and sea snail (*Thais chocolate*) accompanied the appearance of the mussel banks. In 2012, the mussel population continued to expand spatially. In general, the number of benthic species increased with distance from the outfall, varying from 39 to 76 species, and abundance was lower at the outfall area and the highest near the shoreline. Nonetheless, no adverse effects were attributed to the brine discharge on the biota.

6.1.3.3 Taiwan

Two SWRO plants were built in Penghu County. One plant started to operate in 2003 and was expanded in 2006 to a total capacity of 15,500 m^3/day. The second plant was completed in 2008 and at the time of the study (2010–11) was only running with trial runs and monthly maintenance. Brine was discharged through two marine outfalls located ca. 3 km offshore. Temperature at the outfalls was not different from ambient during each survey while salinity was somewhat increased near the outfalls. Zn and Cu concentrations in the brine prior to discharge were relatively higher than other metals and seasonally dependent, possibly a result of variable production. Relatively higher concentrations of Zn and Cu were also observed in the sediments near both outfalls, attributed to brine discharge and perhaps to other anthropogenic sources. Cr and Pb concentrations were also elevated but attributed to local contaminations or background concentrations and not to the brine. Relatively higher concentrations of Zn and Cu were observed in the bivalves, no species given, near (50 m) the outfall, with several incidences of high As levels occurring in certain seasons.

6.1.4 Atlantic Ocean

6.1.4.1 Spain

The Maspalomas II SWRO plant started to operate in 1988 and had a total capacity of 25,000 m^3/day. Brine (S = 69.5, ambient 36.8) was discharged through an outfall 300 m from the shoreline at 4 m depth. In mid-2011, a diffuser was installed at the outfall. The plant's RO membranes were cleaned weekly with sodium metabisulfite (SMBS; $Na_2S_2O_5$, 1600 mg/L) for 45 min, followed by its marine discharge with the brine. Without diffusers, the salinity at 250 m from the discharge point was 39.2 and at 700 m, 37.4. Salinity was always lower than 38 following the installation of the diffusers. Regular brine discharge did not affect the DO_{sat}. Weekly SMBS discharge without diffusers reduced DO_{sat} to <5% for >40 min at 250 m from the discharge point and to 58% for 15 min at 700 m. pH was lower than 6 for almost 20 min at 250 m from the discharge point, and 7.6 for 15 min at 700 m, both lower than expected natural values in seawater (8.0–8.3). Dead individuals of the lizard fish *Synodus synodus* and other soft-bottom fish species

(*Bothus podas*, *Microchirus azevia*, *Trachinus draco*) were observed in an area larger than the area delimited by 10% DO_{sat}. The effects of SMBS discharge were not studied after the installation of the diffuser.

The Las Burras SWRO plant started to operate in 1999 with a total capacity of 25,000 m^3/day. Brine was discharged through a marine outfall 300 m from the shoreline at 7 m depth. Salinity at the outfall ranged from 47 to 50 during the study (2008, 2009) compared to the ambient average of 36.7. Elevated salinities were measured up to 30 m from the outfall in seawater and sediment pore waters. Macrobenthos abundance and species density (number of species per area) were lower at the outfall, but species richness (number of species per number of individuals) was higher. The contributions of fine sands in the sediments and seawater salinity explained 36% and 11%, respectively, of the total variability in macrobenthos assemblages. The effect was localized at the outfall. Meiofaunal assemblage structure was different at the outfall and the abundance lowest, the latter increasing with distance.

6.1.4.2 United States

The Tampa Bay SWRO plant started to operate in 2003, went offline for retrofit in mid-2005, and restarted operations by April 2007. The plant had a total capacity of 95,900 m^3/day and used an open intake for feedwater supply. Brine (S = 54–62) was discharged through an open canal after mixing (1:70 ratio) with cooling waters from the adjacent TECO Big Bend power plant. Surface waters at the discharge area were generally less dense than deeper waters, due to temperatures higher than ambient by 3–4°C. Salinity was near ambient at the intake canal and slightly higher (by 0.5) at the discharge channel during the production periods. An extreme variation in natural salinity was recorded at the vicinity of the plant from 1974 to 2009: salinity ranged from 9.8 to 33.4 (21–30 annual average), with the lowest salinities measured during 2003–05. Benthic macroinvertebrate assemblages were studied from 2002 to 2009 in three areas: 300 m, 1.5 km and 2.5 km from the discharge site. Salinity was similar at the three areas as were the benthic communities during each survey. The biotic responses followed the natural, temporal bay-wide salinity variations; abundance and diversity in all three areas were lower during the first production period with the lowest salinity and highest during the second production period. Diversity was higher at the reference area both during production and no-production periods, with smaller differences during the two production periods. The authors concluded that brine discharge did not impact the benthic macrofauna.

6.2 IN SITU EXPERIMENTS AND LABORATORY BIOASSAYS WITH NATURAL LOCAL SPECIES

In situ experiments and laboratory bioassays test the biotic response to an abiotic stressor such as salinity, temperature, or chemical compound. Although the literature on the effects of salinity and temperature on organisms is extensive, the scope of this section is limited to studies performed with natural local species within the context of desalination. Laboratory bioassays have the disadvantage of being constrained by time and size and of not replicating the natural environmental conditions or the variable exposure of an organism to the brine. However, they do have an added value to in situ studies. Laboratory bioassays can follow the biotic response of an organism exposed to one stressor, under controlled conditions, possibly isolating it from the response to other stressors. However, it is recognized that it is hard to extrapolate the findings of a controlled laboratory experiment to the highly variable in situ environmental conditions. Four in situ experiments and 19 laboratory bioassays were identified from 2001 to 2017. Hypersalinity was the stressor in all studies, with five studies adding temperature, pH, or a chemical as an additional stressor. Marine seagrasses were the most studied organism in 16 out of the 23 studies. Table 6.2 summarizes the details of the studies described in this section and complements the text. The studies were categorized by in situ experiments or laboratory bioassays, the latter organized by organism.

6.2.1 In Situ Experiments

Two in situ experiments were conducted on seagrasses in Spain, one on coral symbionts and one on fish in the Red Sea.

During the pilot stage of the New Channel of Cartagena SWRO (2001–02), brine (S = 70–75) was diverted into a natural seagrass meadow of *Posidonia oceanica*. The plants were exposed for 100 days to 38.6 and 40.1 salinities compared to the ambient 37.6. The exposure was preceded and followed by 426 and 80 days, respectively, of observations at ambient salinity. The meadows declined through a significant decrease in shoot density at both salinities (by 12% and 19%). Surviving shoots had reduced size, lower leaf growth rate, and increased necrotic leaf area compared to control plants.

Plant fragments of the seagrass *Cymodea nodosa* with intact attached rhizome-connected shoots and roots were transplanted in 2009 from well-preserved meadows to two impacted areas near the brine discharge

Table 6.2 Compilation of details of in situ experiments and laboratory bioassays performed with natural local species

Country, area	Date of study	Test organism; response	Stressor	Exposure time (volume)	Exposure concentration (control)	Effect Y/N	Reference
In situ experiments							
Spain, Murcia	2001–02	*Posidonia oceanica* (whole plants); shoot density, growth rate	S (brine)	100 days	38.4, 39.2 (37.6)	Y	Ruiz et al. (2009)
Spain, Alicante	2009	*Cymodea nodosa* (fragments); leaf growth and mortality	S (brine) T (brine)	30 days	38.6–39.9 37.7–38.25 $\Delta T = 0.5–1°C$	Y	Garrote-Moreno et al. (2014)
Saudi Arabia, KAUST, Red Sea	NR	Algal symbiont of the coral *Fungia granulosa;* photosynthetic efficiency and bleaching	S (brine)	29 days	NR (39)	N	van der Merwe et al. (2014a,b)
Saudi Arabia, Gulf	NR	Rabbitfish, metal accumulation	S (brine)	5 days	NR	N	Saeed et al. (2017)
Laboratory studies							
Seagrasses							
Spain, Alicante	NR	*Posidonia oceanica* (whole plants); Shoot and leaf growth and survival	S	15 days (300 L)	25–57 (36.8–38)	Y	Fernández-Torquemada and Sánchez-Lizaso (2005) and Sánchez-Lizaso et al. (2008)

Continued

Table 6.2 Compilation of details of in situ experiments and laboratory bioassays performed with natural local species—cont'd

Country, area	Date of study	Test organism; response	Stressor	Exposure time (volume)	Exposure concentration (control)	Effect Y/N	Reference
Spain, Alicante	2004	*Posidonia oceanica* (fruits); seed germination	S	14 days (NR)	40, 43, 46, 49 (37)	N?	Fernández-Torquemada and Sánchez-Lizaso (2013)
Spain, Alicante	2004	*Posidonia oceanica* (fruits); seed germination and early seedling development	S	50 days (NR)	25–51 (37)	Y	Fernández-Torquemada and Sánchez-Lizaso (2013)
Spain, Murcia	2008	*Posidonia oceanica* (fragments); osmotic potential, leaf growth, shoot survival	S	47 days (500 L)	39, 41, 43 (37)	Y	Sandoval-Gil et al. (2012a)
Spain, Murcia	2008	*Posidonia oceanica* (whole plants); photosynthetic rate, growth	S	47 days (500 L)	39, 41, 43 (37)	Y	Marín-Guirao et al. (2011)
Spain, Murcia	2010	*Posidonia oceanica* (whole plants), osmotic potential physiology survival	S	1, 3 months (500 L)	43 (37 ambient)	Y	Marín-Guirao et al. (2013)
Italy, Ischia	2010	*Posidonia oceanica* (whole plants, germinated seedlings); aquaporin gene expression	S pH	12, 24, 48 h	45 (37) 8.0 (6.0)	Y Y	Serra et al. (2012)

Spain, Almadraba, Alicante	NR	*Cymodocea nodosa*; growth rate and survival	S S and T	10 days (NR)	2–72 (37) 43, 48, 53 (37) 20°C (25°C)	Y Y	Fernández-Torquemada and Sánchez-Lizaso (2011)
Spain, Isla Plana, Murcia	2008	*Cymodocea nodosa* (fragments); osmotic potential	S	47 days (500 L)	39, 41, 43 (37)	Y	Sandoval-Gil et al. (2012a)
Spain, Isla Plana, Murcia	2008	*Cymodocea nodosa* (fragments); photosynthetic rate, respiration	S	47 days (500 L)	39, 41, 43 (37)	Y	Sandoval-Gil et al. (2012b)
Spain, Maspalomas II Canary Islands	ND	*Cymodocea nodosa* (seedlings); survival, growth and necrosis, leaves per shoot	S SMBS S + SMBS	25 days, 40 min weekly exposure to SMBS (NR)	39 (36.8) 100 mg/L 36.8 and 39, and 100 mg/L	Y Y Y	Portillo et al. (2014)
Spain, Almadraba, Alicante	NR	*Zostera noltii*; growth rate and survival	S	10 days (NR)	2–72 (37)	Y	Fernández-Torquemada and Sánchez-Lizaso (2011)
United States, Florida Bay	NR	*Thalassia testudinum*; photosynthetic rate, growth	S	3 days (4 L)	35–70 (35), rapid	Y	Koch et al. (2007)
United States, Florida Bay	NR	*Thalassia testudinum*, *Halodule wrightii*, *Ruppia maritima*; photosynthetic rate, growth, mortality	S	30 days (500 L)	35–70 (35), gradual	N (up to 55–65)	Koch et al. (2007)

Continued

Table 6.2 Compilation of details of in situ experiments and laboratory bioassays performed with natural local species—cont'd

Country, area	Date of study	Test organism; response	Stressor	Exposure time (volume)	Exposure concentration (control)	Effect Y/N	Reference
Australia, Fremantle	NR	*Posidonia australis* (shoots); photosynthethic rate, growth, survival, osmolality	S	6 weeks (200 L)	46, 54 (37)	Y	Cambridge et al. (2017)
Microbial communities and phytoplantkon							
Australia	NR	*Thalassiosira pseudonana, Chaetoceros muelleri,* diatoms; silica structure	S	28 days (NR)	46 (36)	Y	Vars et al. (2013)
Saudi Arabia, KAUST	NR	Dinoflagellate *Symbiodinium microadriaticum,* coral symbiont; cell counts and Chl-a	S	7 days (NR)	25–55	Y	van der Merwe et al. (2014a,b)
Israel	2013	Planktonic microbial community; primary and bacterial productivity, abundance, community structure	S	2 h 11–12 days (1 m^3)	40.9, 45.3 (ambient 39)	Y	Belkin et al. (2015)

Israel	2013–14	Planktonic microbial community; primary and bacterial productivity, abundance, community structure	AS alone Fe alone S + AS + Fe	2 h 11–12 days ($1\,m^3$)	0.2 mg P per L 1 mg Fe per L 44.5, 0.2, 1	Y Y	Belkin et al. (2017)
Israel	2015	Benthic heterotrophic bacteria; abundance, community structure, activity	S	48 h (3 L)	39.8, 41, 46.8 (39)	Y	Frank et al. (2017)
Other organisms							
United States, Carlsbad, CA	ND	18 organisms; survival and behavior	S (brine)	5.5 months (NR)	36 (33.5)	N	Le Page (2005) and Voutchkov (2007)
United States, Carlsbad, CA	ND	Purple sea urchin (*Stronglyocentroutus purpuratus*), the sand dollar (*Dendraster excentricus*), and the red abalone (*Haliotis rufescens*); survival and behavior	S (brine)	19 days (NR)	37–40 (33.5)	N	Le Page (2005) and Voutchkov (2007)
Spain, Alicante	ND	Mysid crustacean *Leptomysis posidoniae*, sea urchin *Paracentrotus lividus;* Mortality	S S and T	15 days (300 L)	NR	Y	Sánchez-Lizaso et al. (2008)

Continued

Table 6.2 Compilation of details of in situ experiments and laboratory bioassays performed with natural local species—cont'd

Country, area	Date of study	Test organism; response	Stressor	Exposure time (volume)	Exposure concentration (control)	Effect Y/N	Reference
Australia, Upper Spencer Gulf	2007	Giant cuttlefish, *Sepia apama*, (eggs); hatching, survival, growth	S	3 months (NR)	40, 45, 50, 55 (39)	Y	Dupavillon and Gillanders (2009)
Spain, Maspalomas II Canary Islands	ND	Fish *Synodus synodus*; toxicity	SMBS	40 min (NR)	25, 50 mg/L	Y	Portillo et al. (2014)
United States, California	ND	Japanese medaka (*Oryzias latipes*), fish embryos; hatch, survival and deformities	S	NR (10 mL)	17, 39, 42, 46, 50, 56, 70 (35)	Y	Kupsco et al. (2017)

?, inconclusive; *AS*, Polyphosphonate antiscalant; *Chl-a*, chlorophyll-a; *DO*, dissolved oxygen; *Fe*, iron salt coagulant; *N*, no; *NR*, not reported; *S*, salinity; *SMBS*, sodium metabisulfite; *T*, temperature; *Y*, yes.

of the Alicante I and II SWRO plants and to two control areas, at 10 m depth. Four one-month experiments were performed. The number of leaves per shoot and leaf growth of vertical shoots at the impacted localities were lower than the control during the four experiments. The lowest leaf growth and the highest shoot mortality (83%) were detected in July and August, when seawater temperatures were higher. Significant higher mortality was found at the impact area with the higher salinities. The number of leaves per shoot and leaf growth of the horizontal shoots were not affected by salinity, therefore the ratio of horizontal to vertical shoots increased at the impact stations. Salinity tolerance was lower than those found in laboratory bioassays.

Corals (*Fungia granulosa*) from a reference site (S = 39) were transplanted to the KAUST SWRO discharge screen and to the bottom along a 25 m transect. Salinity at the discharge screen was 49.4 ± 2.0 and lower at the other stations, with an average of 41.3 ± 0.7. Temperature was also higher at the discharge screen, but no significant differences were found in DO concentration. After 29 days of exposure, brine did not impact the photosynthetic efficiency of the algal symbiont and no bleaching was detected.

No metal accumulation was detected in liver and muscle of the Rabbitfish (*Siganus rivulatus*) held for 5 days in cages at the intake and brine discharge at the Jeddah MSF and SWRO plants.

6.2.2 Laboratory Bioassays on Seagrasses (Mediterranean Sea (Spain, Italy); Atlantic Ocean, Gulf of Mexico (United States); and Indian Ocean (Australia))

6.2.2.1 Posidonia oceanica

The endemic seagrass *P. oceanica* constitutes one of the most ecologically important shallow marine habitats in the Mediterranean Sea. In situ observations have shown this species to be very sensitive to hypersalinity, their meadows decreasing following exposure to desalination brine (see Sections 6.2 and 6.2.1). Consequently, the response of *P. oceanica* to hypersalinity was studied in a series of laboratory experiments performed in Spain and Italy from 2004 to 2010. Whole plants, fragments (i.e., shoots connected to a basal rhizome with the root system intact, Fig. 6.2) or fruits were collected, brought to the laboratory, and exposed to different salinities in small to large 500 L tanks. Germination, growth, physiological and photosynthetic responses, and mortality rate were followed. The general conclusion was that *P. oceanica* is very sensitive to hypersalinity even at small increases above ambient. The effects were highly dependent on the mode, length, and intensity of exposure. At some instances, plants were able to recover when returned to ambient salinity. Following is the description of the specific studies.

Seed germination experiments were conducted for 14 days at 40–49 salinities and compared to a control treatment (S=37). The highest seed germination (90%) was reached at the control, but it was not statistically different from the other treatments (63%–80%). Mortality was lowest for the control treatment (13%), increasing considerably at salinities above 46 (53%–73%), but not statistically significant. Early seedling development experiments were conducted for 50 days in 14 treatments with salinities ranging from 25 to 51. The maximum leaf and root lengths and the number of produced leaves were larger at the control and low salinity exposure compared to the hypersalinity treatments. While the highest seedling survival (90%) and development was observed at the control treatment, mortality was higher at low salinity (S=25–35, 50%–80%) than at high salinities (20%–40%).

Shoot growth and survival of *P. oceanica* plants decreased following 15 days exposure to salinities from 25 to 57 in 300 L tanks. Maximum leaf growth occurred between 25 and 39 salinities, while considerable plant mortality occurred at salinities above 42 and below 29, with 100% mortality at 50. Plants were able to regain their original growth rate when returned to ambient seawater following exposure to salinities from 39 to 46.

Large *P. oceanica* fragments and whole plants were exposed for 47 days to salinities of 39, 41, and 43 (ambient S=37) in 500 L aquaria. In the fragments, osmotic potential decreased while the concentrations of soluble sugars and some amino acids in the leaves increased following exposure to increased salinity, suggesting the activation of osmoregulatory processes. Osmotic adjustments probably interfered with leaf growth and shoot survival under hypersaline stress. In the whole plants, both net and gross photosynthetic rates under hypersaline conditions were significantly reduced. The photosynthetic apparatus to capture and process light was not affected, but dark respiration rates increased substantially, suggesting that the respiratory demands of the osmoregulatory process were responsible for the decrease in photosynthetic rates. Leaf carbon balance was considerably impaired, and leaf growth rates decreased as salinity increased. No significant differences were found in the percentage of net shoot change, but mean values were negative at salinity levels of 41 and 43. Whole plants were also exposed in similar aquaria to salinity of 43 for 1 and 3 months, followed by a 1-month recovery at ambient salinity (37). One-month saline-stressed plants exhibited sublethal effects (leaf cell turgor pressure reduction, loss of ionic equilibrium, and decreased leaf growth), changes in photoprotective mechanisms, and increased concentrations of organic osmolytes in leaves. The plants recovered after 1 month at ambient salinity. A longer exposure of 3 months caused severe physiological stress: excessive ionic exclusion capacity, increased leaf cell turgor, reduced plant carbon balance, increased leaf aging and leaf decay, and increased plant mortality. The long-term saline-stressed plants were not able to recover after 1 month at ambient salinity.

Mature plants of *P. oceanica* and young leaves from germinated seedlings were exposed to hypersalinity stress of 45 (37, control) for 12, 24, and 48 h. Gene expression of two aquaporin (PIP1and PIP2) water channels, essential in regulating water exchange, was examined. In adult plants, PIP2 showed a twofold increase in expression levels at all exposure times, while PIP1 was significantly up-regulated only after 24 h. In young *P. oceanica* leaves, salinity variations did not induce significant changes in both PIP1 and PIP2 expression level, suggesting the inability of the seedlings to deal with this stressful condition. Similar exposure of mature plants to low pH (6 compared to control 8) induced a twofold up-regulation of PIP1 after 24 and 48 h exposure and of PIP2 at all times, slightly higher at 24 h. In young leaves, PIP2 was almost twofold up-regulated at all exposure times, while PIP1 expression levels increased significantly only after 12 and 24 h.

6.2.2.2 Other seagrasses

Similar laboratory studies were performed with seagrasses found in the Mediterranean other than *P. oceanica* and with seagrasses from the Atlantic-Caribbean and the Indian Ocean. The general conclusion was that seagrasses may be sensitive or tolerant to hypersalinity, mainly depending on the species but also on the mode, length, and extent of the exposure. Following are the summaries of the studies identified.

Leaves of large fragments of the seagrass *Cymodocea nodosa* (Mediterranean Sea) exposed to hypersalinity (S = 39, 41, 43, control 37) for 47 days in 500 L aquaria reduced their turgor pressure to prevent dehydration with no activation of osmoregulatory processes. *C. nodosa* showed a more efficient physiological capacity to maintain plant performance under hypersaline stress than *P. oceani*ca. Net photosynthesis rate declined slightly (12%–17%) in all hypersaline conditions. At intermediate salinity (39, 41) the rate decline increased respiratory losses, but the negative effects on leaf carbon balance were offset by an improved efficiency to absorb light. Conversely, inhibition of gross photosynthesis at S = 43 reduced net photosynthesis. The respiration rate was limited in order to facilitate a positive carbon balance and shoot survival, although vitality would probably be reduced if such metabolic alterations persisted.

Forty-five-day-old seedlings of the seagrass *C. nodosa* from the Canary Islands were exposed to hypersalinity (S = 39, ambient 36.8), to SMBS (100 mg/L), and to both stressors combined for 25 days. The exposure to SMBS was done weekly for 40 min, simulating the desalination plant operations (see Section 6.2). Salinity alone did not affect seedling survival while SMBS addition at both salinities decreased the survival. Salinity decreased the leaf elongation rate and increased necrotic leaf surface with a larger impact with additional SMBS exposure. Exposure to salinity or SMBS alone decreased the mean total leaf surface area of shoots, with no synergistic effect when both stressors were present. No effect was detected on the number of leaves per shoot.

Short term (10 days) laboratory experiments were performed on the seagrasses *Zostera noltii* and *C. nodosa* from the Mediterranean Sea exposed to salinities from 2 to 72 (control 37). Shoot growth rate of *Z. noltii* was not affected at the low salinities but considerably affected at salinities above 41. Mortality was 17% at S = 43, 50% at S = 50, and 100% at S ≥ 57. Maximal shoot growth rate in *C. nodosa* was observed from 30 to 39 salinity, with a significant reduction at salinities higher than 41 or lower than 16. Mortality was minimal at the control, increasing toward low and high salinities, in particular at S > 50. All plants died when exposed to salinities of 2 and 57. The magnitude of the responses was larger in the summer than in the winter. Exposure of *C. nodosa* to higher temperature (25°C, control 20°C) in conjunction with hypersalinity (43, 48, 53, control 37) increased shoot growth rate except at S = 53. Temperature did not affect mortality of shoots. *C. nodosa* seemed to acclimate to gradual increases in salinity (2.5 per day) compared to a sudden change in salinity.

The tropical seagrass, *Thalassia testudinum* is the dominant bed forming seagrass in Florida Bay and in the wider Atlantic-Caribbean region. *T. testudinum* was exposed to hypersalinity (35–70, control 35) in short term (3 days) experiments conducted in 4 L containers. Salinity was raised rapidly to the experimental value. The photosynthetic rate decreased with increased salinity, statistically significant at salinities higher than 60. Leaf growth rates decreased with increase in salinity above 45 (by up to 42%) reaching a low value at S > 50, indicating stress. In a long-term experiment (30 days), *T. testudinum* and the common *Halodule wrightii* and *Ruppia maritima* were exposed to salinities from 35 to 70 in 500 L aquaria. The salinity was increased gradually to the final salinity at a rate of 1.0 per day. The three seagrasses were highly tolerant of hypersaline conditions during the gradual increase of salinity, probably due to their ability to osmoregulate. The stress indicators (shoot decline, growth rates, and quantum yields of photosynthesis) demonstrated tolerance up to salinity of 55–65. Hypersalinity did not affect shoot mortality. However, when salinity increased rapidly, without a slow osmotic adjustment period, salinity tolerance dropped to about 45 for *T. testudinum*.

Shoots of the seagrass *Posidonia australis* from Australia were exposed for 6 weeks to salinities 46 and 54 (control 37), in 200 L tanks. Hypersalinity increased plant morbidity and mortality, reduced growth of roots, induced highly negative water potentials with no change in turgor pressure, increased the concentration of osmoregulators, and reduced concentrations of K and Ca ions in the leafs. Plants survived exposures of 2–4 weeks at S = 54. After 6 weeks, mortality was 33% and 69% at salinities of 46 and 54, respectively. The photosynthetic capacity of *P. australis* was maintained with little change for several weeks at both salinities and inhibited after 4 weeks of exposure at S = 54. After 6 weeks, the photosynthetic markers were reduced in both hypersaline treatments.

6.2.3 Laboratory Bioassays With Organisms Other Than Seagrasses

Salinity tolerance tests were performed within the EIA for the Carlsbad SWRO plant (CA, United States) using three local species: purple sea urchin (*Stronglyocentroutus purpuratus*), sand dollar (*Dendraster excentricus*), and red abalone (*Haliotis rufescens*). The organisms were exposed for 19 days to salinities 37–40 (ambient 33.5) expected at the edge of the zone of initial dilution. The three test species had a survival rate of 100%, with individuals moving and feeding normally, showing salinity tolerance to the desalination plant discharge. During a long-term exposure (5.5 months) at S = 36, specimens of 18 species remained healthy, with good appearance, no changes in coloration or development of marks or lesions. Among the tested organisms were the barred sand bass, California halibut, red sea urchin, and green abalone.

The mysid *Leptomysis posidoniae* and the sea urchin *Paracentrotus lividus*, organisms frequently found in *P. oceanica* meadows, were exposed to hypersalinity in 300 L aquaria. Mortality of these two species increased at salinities over 40.5. Temperature as an additional stressor further increased the mortality rate.

Eggs from the giant Australian cuttlefish (*Sepia apama*) were exposed in 2007 to hypersalinity (S = 40, 45, 50, 55, ambient 39) for 3 months. Hatching success was similar (ca. 95%) at the control and S = 40 treatments and decreased to 70% at S = 45. Total mortality of eggs occurred at S ≥ 50. Smaller mantle length and weight were found in individuals hatched in the laboratory at S ≥ 45, while all the laboratory individuals were significantly smaller and lighter than specimens collected from the field.

The silica structure was studied on the diatoms *Thalassiosira pseudonana and Chaetoceros muelleri* grown in the laboratory under ambient conditions (S = 36) and hypersalinity (S = 46) for 28 days. Under hypersalinity, only *C. muelleri* produced more condensed, hydrophobic, and rigid silica structures and/or organic material linked to the structure, and more extracellular polymeric substances. Both species showed the appearance of a new nitrogenous purine-type compound.

Symbiodinium microadriaticum, the dinoflagellate symbiont of the coral *Fungia granulosa* was cultured under salinities ranging from 25 to 55. Cell counts and Chl-a concentrations showed that the exponential growth phase was reached after 4 days at salinities from 30 to 50. The quickest cell proliferation was observed at S = 35. Growth rates were inhibited in cells exposed to salinities 25 and 55 for 7 days.

The fish *Synodus synodus* revealed a high sensitivity to short-term exposure (40 min) to sodium metabisulfite (SMBS) used to clean the RO membranes (see Section 6.2). Total mortality occurred within 10 min at SMBS concentrations equal to or higher than 50 mg/L. SMBS presence reduced DO_{sat} to <10%, and lowered pH to <6.6. No mortality occurred following exposure to 25 mg/L SMBS, when DO_{sat} was 51% and seawater pH 7.2.

A series of mesocosm experiments were conducted on natural assemblages of planktonic microbial populations from the Mediterranean Sea in 2013–14. The populations were exposed to hypersalinity (S = 40.9, 45.3, ambient 39), iron salt (coagulant, 1.0 mg Fe per L), and polyphosphonate (antiscalant, 0.2 mg P per L) in 1000 L enclosures. Rapid exposure (2 h) to hypersalinity (S = 45.3) reduced primary productivity and algal biomass while heterotrophic bacterial productivity increased. Rapid exposure to the coagulant changed the composition of the bacterial communities significantly, increased the heterotrophic productivity, and reduced primary productivity. Exposure to the antiscalant alone relieved the phosphorous stress of the community. Rapid combined exposure to all stressors (coagulant, antiscalant, and hypersalinity) resulted in synergistic effects reflected by increased productivity of both primary and bacterial producers (by 100% and 50%, respectively). Longer exposure (10–12 days) to hypersalinity elicited responses mostly in the composition and structure of bacterial and eukaryotic communities accompanied by functional changes of enhanced primary and heterotrophic bacterial productivity (by 100%–150% and by up to 200%, respectively). The response was seasonally dependent, with a more distinct impact on the summer communities. In general, the relative abundance of large cyanobacteria increased and of the small cyanobacteria decreased under hypersalinity conditions. Some classes of heterotrophic bacteria increased while other decreased. Longer exposure to iron salt caused a significant decline in autotrophic biomass and in assimilation number while longer exposure of phosphonate increased the ratio of bacterial production to abundance. Significant compositional shifts occurred following exposure to all stressors combined: reduced function (photosynthetic rates) and biomass of primary producers by 50% and increased heterotrophic bacterial activity and communities by 50%.

Hypersalinity effects on the embryonic development of the euryhaline model fish Japanese medaka (*Oryzias latipes*) were followed after exposure to artificial seawater (salinities 17, 35, 42, 56, 70) and RO brine (salinities 39, 42, 46, 50) in the United States. No significant difference was observed between the responses to artificial seawater and to brine. Hypersalinity significantly decreased percent hatch, down to 22% at S = 70. All embryos at this salinity were dead immediately posthatch. Deformities increased to 50% at S = 56 and swim bladder inflation decreased to 46% in treatments with S = 46. Three-day posthatch embryos survival was 78% at S = 42 and 0% at S = 70. Exposure to hypersalinity increased the time to hatch following egg fertilization.

Natural benthic heterotrophic bacteria attached to marine sediments from the Mediterranean Sea were exposed for 48 h to hypersalinity (39.8, 41, and 46.8, control 39) in 3 L cylindrical tubes. In the summer, abundance was reduced by 60% at the two highest salinities, and bacterial cell-specific activity increased. Bacterial abundance was not affected in the winter. Hypersalinity did not affect bacterial community structure, bacterial production, nor respiration in both seasons.

6.3 TOXICITY TESTING

Toxicity tests compare the response of an organism exposed to a specific chemical at various concentrations to the response of the same organisms unexposed to the chemical, called the control. Similarity, whole effluent toxicity (WET) tests measure the organism's response to the aggregate toxic effect from all pollutants present in an effluent (see Chapter 7). Standard toxicity and WET tests use model organisms and measure their ability to survive, grow, and reproduce. These tests are used to derive specific concentrations such as: lowest observed effect concentrations (LOEC); no observed effect concentrations (NOEC); EC_{50}, concentration that causes an effect on 50% of the population; IC_{50}, concentration that causes an inhibition of growth of 50% in unicellular algae bioassay. The test results can be used to calculate dilution of the effluents needed to protect the biotic population to a predetermined level. Toxicity and WET tests are usually performed within an EIA (see Chapter 7) prior to the construction of a desalination plant and presented in hard-to-find internal reports. Their results are not easily integrated in research but are used for regulatory purposes. Below are nine studies that exemplify these tests.

The brine shrimp Artemia (*Artemia franciscana*) was exposed to seawater collected at the intake and discharge sites from four desalination plants in the KSA: Jubail in the Gulf, and Jeddah, Haql, and Shuqaiq in the Red Sea. The hatching rate of Artemia cysts was $\geq$95% in all treatments. No larval mortality occurred 24 and 48 h after hatching.

Toxicity tests were conducted within the EIA and monitoring studies for the Adelaide SWRO desalination plant, from 2009 to 2013. Common marine species native to South Australia (Polychaeta, crustacean, plant, fish and phytoplankton) were exposed for 24 h and 7 days to different concentrations of the following solutions: brine, brine treated with chlorine and neutralized with sodium metabisulfite, permeate treated with various chemicals simulating the desalination process, and seawater treated with flocculants. Typically, the results showed a progressively increasing negative response across the concentration series. The calculated safe brine dilution required for the protection of 99% of the species at the discharge site ranged from 10 to 20. The actual planned brine dilution following discharge was 50. Only the neutralized permeate treated with NaOH required up to a 2500 dilution. Consequently, the marine discharge of this solution was prohibited and disposal took place via the sewer system.

Toxicity test were performed in 2015 with effluents from the Southern SWRO plan, Binningup, Australia, that started operating in 2012. The solutions used were MF and RO effluents containing antiscalant and neutralized chemicals used to clean the membranes (detergent, chlorine, citric acid, sodium metabisulfite, sulfuric acid, and sodium hydroxide) at various dilutions. EC_{10} from 15% to 25% desalination effluent were found for the larval development of mollusk, fish, and copepod. The copepod reproduction bioassay resulted in an EC_{10} of 32% desalination effluent. The Microtox bioassay resulted in an EC_{10} of 79% desalination effluent while the microalgae growth and macroalgae zoospore germination were unaffected by the desalination effluent.

WET tests were performed in 2006–07 within the framework of an EIA for the proposed Point Lowly SWRO desalination plant, South West Australia. The test species were exposed to salinities ranging from ambient (36.3) to 52 and to the undiluted brine (S = 78). The tests included: 72-h growth rate in the microalgae diatom *Nitzschia closterium* (chronic), 72-h larval development test on the sea urchin, *Heliocidaris tuberculate* (subchronic), 96-h survival test on the yellowtail kingfish *Seriola lalandi* and on the prawn *Penaeus monodon* (acute), 72-h germination test on the macroalgae *Hormisira banksia* (chronic), and 48-h larval development test on the oyster *Saccostera commercialis* (subchronic). The most sensitive species was the sea urchin (NOEC at 4% effluent, S = 38) known to be a stenohaline osmoconformer and the least sensitive were the prawn and the macroalgae (NOEC at 17% effluent, S = 43). Toxicity was attributed to salinity for all organisms except for the microalgae that 30% of the toxicity was attributed to the polyphosphonate antiscalant. The brine dilutions to protect 95% and 99% of the species were estimated to be 1:60 and 1:80 respectively.

Toxicity tests were performed within the extensive EIA for the Carlsbad SWRO plant, Southern California, United States. Brine was obtained from a small RO unit at the Encina Power Station, located next to the planned desalination plant. The tests performed with diluted brine were: giant kelp (*Macrocystis pyrifera*) 48-h chronic germination and growth; Pacific topsmelt fish (*Atherinops affinis*) 7-days chronic survival and growth; and red abalone (*Haliotis ruefescens*) embryonic development over 48 h postfertilization. No effects were found in these cases. Chronic toxicity threshold for the abalone was salinity higher than 40 and for the topsmelt, salinity higher than 50. The same toxicity tests, using the Pacific topsmelt, giant kelp, and red abalone, were performed three times in 2017, following the start of plant operations of the Carlsbad SWRO plant in 2016 (total capacity of 189,500 m^3/day). A 24-h composite discharge sample collected at the plant was diluted with seawater to test concentrations: 2.5%, 5%, 6%, 10%, and 15% of brine. The maximal salinity the organisms were exposed to was 37.9–38.3, compared to the ambient 33.5. NOEC was 15% of the discharge for the Pacific topsmelt and the giant kelp and 10% of the discharge for the red abalone. There were no adverse effects observed. All test results passed and complied with the discharge permit.

6.4 CRITICAL EVALUATION AND INTEGRATION OF RESULTS

Ideally, the data collected on the actual impacts of desalination could be integrated following their adjustment to reflect the different conditions under which the studies were performed. A general model would then be derived, similar to the one depicted in Fig. 6.5, and applied to any discharge site.

In reality, it is impossible, at this stage, to propose such a model due to the small number of studies conducted under highly variable conditions. Several factors contribute to the scant number of publications:

(1) Desalination arose out of urgent necessity for freshwater, with pressing timelines, which resulted in omitting EIA studies and environmental monitoring.

(2) Initial lack of awareness to environmental impact may have deemed studies unnecessary. The belief that "if properly engineered, desalination will not affect the environment" is prevalent even today.

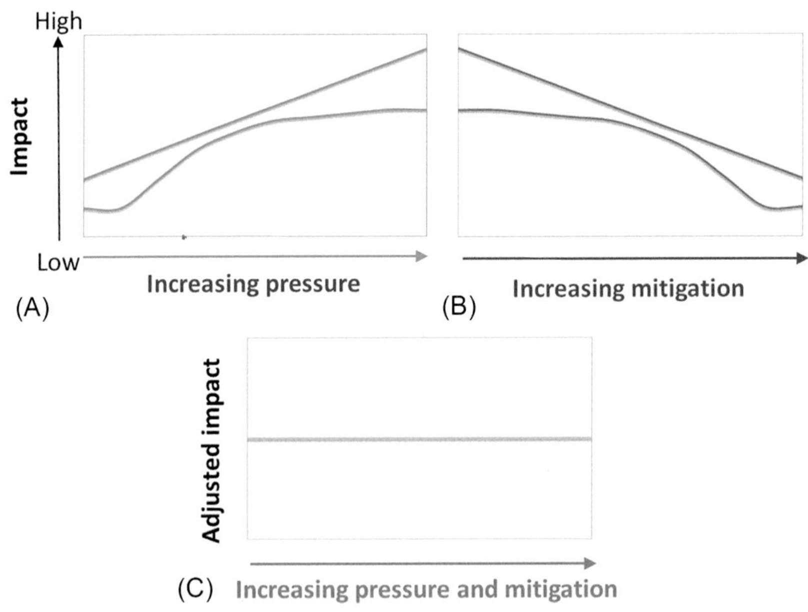

Fig. 6.5 Schematic representation of a generalized stressor-impact relationship: (A) Increasing stress (such as plant size, salinity, temperature, concentration of chemicals in brine), (B) Increasing mitigation (such as brine dilution, distance from coast, distance from outfall, reduction of brine volume) and (C) Adjusted to site-specific conditions.

(3) There may be a bias in the publication effort toward studies that show an observable impact, under the erroneous assumption that no impact is of no interest to the community.

(4) The peer-reviewed journals still consider monitoring studies unworthy of publication. Moreover, monitoring is usually conducted by commercial companies, with no incentive or permission to publish the results.

(5) Specific compliance monitoring (see Chapter 7) is limited in scope, and the data obtained is mainly designated to ensure regulation conformity and not for scientific research.

Further hindrance to a general description of the impacts of desalination, within in situ studies, stems from the lack of long-term environmental records, which are necessary to factor out the natural temporal variability. Many of the studies do not report important details (Tables 6.1 and 6.2), some present incomplete interpretations of the results, while others display potentially erroneous data that need careful screening. In laboratory bioassays, extrapolation of the results to the highly variable exposure to multiple stressors in the real environment is very complex.

Nevertheless, it was possible to reach some general conclusions, point out problems with specific studies and map the details needed in a study to facilitate integration of the results in the future. The most prevalent parameter measured in situ was salinity, appearing in 34 studies. It was followed by temperature (21 studies), benthic communities (21 studies), and metals in sediments (15 studies) (Table 6.1). Taking into account all parameters measured, some appearing in most studies, 43% were reported to be affected by the brine discharge, 6% were inconclusive while the rest (51%) were not affected by the brine discharge. In contrast, 85% of the laboratory bioassays reported some effect while 3% were inconclusive. The differences between in situ studies and laboratory bioassays may be due to a more extreme exposure to a stressor in the latter.

Following is a critical evaluation of the results, organized by stressor. Tables 6.1 and 6.2 complement the text.

6.4.1 Salinity

Salinity was the most common stressor studied in situ and in laboratory bioassays. As natural salinity differs among areas it is more appropriate to compare salinity changes (in percentage) from the ambient salinity to factor out this natural variability.

Not surprisingly, the general conclusion of the in situ studies was that brine discharge increased seawater salinity, an abiotic impact. Seawater salinity showed that brine from SWRO plants dispersed near the bottom while brine from thermal desalination plants or from SWRO plants co-discharging with cooling water from power plants dispersed mostly near the surface. Salinity was higher than ambient mostly by 1%–7%, but 15% and 30% increases over ambient were measured. No increase of salinity was detected in one out of the 34 studies and in one the result was inconclusive. In the former (Gold Coast, Australia), the plant was operating at low capacity and discharging the brine through a diffuser. In the latter (Barka, Oman) the brine was mixed with cooling waters from an adjacent power plant and discharged through four pipelines. SWRO increased seawater salinity more than thermal desalination, probably due to the lower recovery rates in the latter, leading to lower brine salinity.

The extent of the impact was highly site-specific, depending on the capacity of the plant, the desalination process, and the mode of brine disposal. Increased salinity was detected up to 5 km from the discharge site, but most studies reported an effect within tens to a few hundred meters. As expected, large plants discharging the brine with low initial dilution had a larger impact on salinity. For example, hypersalinity was measured for a few kilometers from the outfall of the San Pedro del Pinatar SWRO plant (Spain) prior to the installation of a diffuser while no increase in salinity was detected following its installation.

In situ biotic effects on benthic communities were surveyed in 21 studies. Twelve reported on effects detected at the outfall and up to 600 m from it and attributed them to hypersalinity. The effects included reduced abundance, changes in diversity and community structure, and promotion of salinity tolerant organisms, in particular from the Polychaeta family. Benthic communities recovered following the removal of the salinity stress, as in San Pedro del Pinatar (Spain) following the installation of a diffuser at the outfall. Inconclusive results at five locations stem from a short-term time series making it impossible to account for natural variability (Palmachim and Soreq, Israel) and from reporting changes but not attributing them directly to desalination (La Chimba, Chile; Gold Coast, Australia). Out of the four studies that reported no effect on the benthic communities, one defined the impacted area too far from the outfall (3 km) missing perhaps a possible effect at a closer location (Perth, Australia), and three looked at macrobenthos invertebrates (Blanes, Spain; Adelaide, Australia; Tampa Bay, FL, United States), that may be less sensitive than the infauna.

Hypersalinity was the main stressor used in laboratory bioassays to simulate the brine's impact on biota. Different parts of seagrasses (whole plants, fragments, fruits, seeds; Fig. 6.2) were exposed for 12 h to 3 months, in small to 500 L enclosures, to salinities ranging from 5% higher to twice as high as ambient. While effects were detected in 15 out of 17 studies conducted, most occurred following exposure to very high salinities that are not encountered in situ. Even so, brine was shown to affect seagrass meadows in the Mediterranean Sea exposed to hypersalinity as low as 2% higher than ambient. The reasons for this inconsistency may be the limited exposure time and volume in the laboratory experiment compared to the in situ conditions, the part of the organism exposed, the mode of exposure (rapid or gradual increase in the laboratory as opposed to variable exposure in the field), and the additional components present in the brine and not simulated in the laboratory. Additional obstacles for the derivation of a general threshold of salinity tolerance for a specific seagrass were the different responses studied and the possibility of genetically different populations, for example *C. nodosa* from the Mediterranean Sea or the Atlantic Ocean.

Hypersalinity was shown to affect the pelagic microbial populations in the five laboratory bioassays. In situ, two studies reported similar effects but to a lesser degree (Ashqelon, Israel), and one study showed no affect (KAUST, KSA).

6.4.2 Temperature

Temperature at the outfalls of SWRO plants was higher than ambient by 0.5–1°C. As expected, temperatures were much higher than ambient (up to 10°C higher) at the brine discharge site of thermal desalination plants and when SWRO brine was co-discharged with cooling waters from power plants. The dispersal of temperature followed that of the salinity, as explained in Section 6.4.1. Most in situ studies did not address the possible biotic effects of increased temperature, only salinity. This may be due to the fact that most biotic studies were performed at SWRO outfalls in which the main stressor was salinity, with only a slight increase in temperature. In the laboratory studies, temperature elevation, combined with hypersalinity, reduced growth rate and survival of *C. nodosa* and increased mortality of a mysid crustacean and a sea urchin from the Mediterranean Sea compared to the exposure to hypersalinity alone.

6.4.3 Metals in Sediments

The concentrations of metals in sediments were followed in 15 in situ studies, 6 at discharge sites of thermal desalination plants, and 9 off SWRO plants. An increase of metal concentrations, in particular Cu, was detected at all MSF outfalls, probably originating from metal corrosion in the plant. Metal concentrations were natural off the SWRO's plant discharges except for two cases: Ras Tanajib (KSA) and Pengu (Taiwan). Some studies failed to adjust the results to reflect the natural concentration variability as a function of composition and grain size, so the results should be further reviewed.

To summarize, details matter. Important details, necessary for the evaluation and integration of the results are often not reported, as shown in Tables 6.1 and 6.2. The community should aim for a standardized description of the study site and the experimental set up. For in situ studies those should include, at least, the following:

- Desalination plant: location, process, pretreatment and post-treatment, start date of operations, total capacity, actual capacity during the study, intake and discharge systems, brine composition (salinity, temperature, chemicals), any changes in the process or inactivity periods that occurred during the years of operation.
- Intake and discharge sites: all available data on the natural environment relevant to the study, to serve as reference values.
- Study: date performed, plant operations during the study, detailed sampling procedure including station location, distance from discharge sites, range of exposures to a specific stressor (as percentage over ambient values), reference stations.

Laboratory bioassays should, in addition to the applicable details described herein, report also on the test organism identity and developmental stage, stressor, total volume of experiment, time and mode of exposure, range of exposure (absolute values and percentage from the control), conditions of the control treatment, number of replicates, and comparison to possible in situ conditions.

REFERENCES

Abdul Azis, P.K., Al-Tisan, I.A., Daili, M.A., Green, T.N., Dalvi, A.G.I., Javeed, M.A., 2003. Chlorophyll and plankton of the Gulf coastal waters of Saudi Arabia bordering a desalination plant. Desalination 154, 291–302.

Abdul-Wahab, S.A., Jupp, B.P., 2009. Levels of heavy metals in subtidal sediments in the vicinity of thermal power/desalination plants: a case study. Desalination 244, 261–282.

Alharbi, O.A., Phillips, M.R., Williams, A.T., Gheith, A.M., Bantan, R.A., Rasul, N.M., 2012. Desalination impacts on the coastal environment: Ash Shuqayq, Saudi Arabia. Sci. Total Environ. 421–422, 163–172.

Alharbi, T., Alfaifi, H., Almadani, S.A., El-Sorogy, A., 2017. Spatial distribution and metal contamination in the coastal sediments of Al-Khafji area, Arabian Gulf, Saudi Arabia. Environ. Monit. Assess. 189, 634.

Al-Said, T., Al-Ghunaim, A., Subba Rao, D.V., Al-Yamani, F., Al-Rifaie, K., Al-Baz, A., 2017. Salinity-driven decadal changes in phytoplankton community in the NW Arabian Gulf of Kuwait. Environ. Monit. Assess. 189, 268.

Alshahri, F., 2017. Heavy metal contamination in sand and sediments near to disposal site of reject brine from desalination plant, Arabian Gulf: assessment of environmental pollution. Environ. Sci. Pollut. Res. 24, 1821–1831.

Al-Yamani, F., Yamamoto, T., Al-Said, T., Alghunaim, A., 2017. Dynamic hydrographic variations in northwestern Arabian Gulf over the past three decades: temporal shifts and trends derived from long-term monitoring data. Mar. Pollut. Bull. 122, 488–499.

Anon, 2017. Southern seawater desalination plant. Marine Environment Monitoring Annual Report 17 January 2016–16 January 2017, Prepared by Water Corporation, Australia.

Ayala, V., Kildea, T., Artal, J., 2015. Adelaide Desalinatino Plant-Environmental Impact Studies. The International Desalination Association World Congress on Desalination and Water Reuse 2015, San Diego, CA, USA: IDAWC15-Ayala_51444.

Begher Nabavi, S.M., Miri, M., Doustshenas, B., Safahieh, A.R., 2013. Effects of a brine discharge over bottom polychaeta community structure in Chabahar bay. J. Life Sci. 7, 302–307.

Belatoui, A., Bouabessalam, H., Hacene, O.R., De-La-Ossa-Carretero, J.A., Martinez-Garcia, E., Sanchez-Lizaso, J.L., 2017. Environmental effects of brine discharge from two desalination plants in Algeria (South Western Mediterranean). Desalin. Water Treat. 76, 311–318.

Belkacem, Y., Benfares, R., Houma Bachari, F., 2016. Potential impacts of discharges from seawater reverse osmosis on Algeria marine environment. J. Environ. Sci. Eng. B5, 131–138.

Belkacem, Y., Benfares, R., Adem, A., Houma Bachari, F., 2017. Evaluation of the impact of the desalination plant on the marine environment: case study in Algeria. Larhyss J. 30, 317–331.

Belkin, N., Rahav, E., Elifantz, H., Kress, N., Berman-Frank, I., 2015. Enhanced salinities, as a proxy of seawater desalination discharges, impact coastal microbial communities of the eastern Mediterranean Sea. Environ. Microbiol. 17, 4105–4120.

Belkin, N., Rahav, E., Elifantz, H., Kress, N., Berman-Frank, I., 2017. The effect of coagulants and antiscalants discharged with seawater desalination brines on coastal microbial communities: a laboratory and in situ study from the southeastern Mediterranean. Water Res. 110, 321–331.

Benaissa, M., Rouane-Hacene, O., Boutiba, Z., Guibbolini-Sabatier, M.E., Faverney, C.R.-D., 2017. Ecotoxicological impact assessment of the brine discharges from a desalination plant in the marine waters of the Algerian west coast, using a multibiomarker approach in a limpet, Patella rustica. Environ. Sci. Pollut. Res. 24, 24521–24532.

Bonnelye, V., Chapman, D., Heiner, T., Ferguson, M., Vollprecht, R., Chidgzey, L., 2017. The Perth Seawater Desalination Plant: 10 Years on…. The International Desalination Association World Congress on Desalination and Water Reuse 2017, Sao Paulo, Brazil IDA17WC-57935.

Cambridge, M.L., Zavala-Perez, A., Cawthray, G.R., Mondon, J., Kendrick, G.A., 2017. Effects of high salinity from desalination brine on growth, photosynthesis, water relations and osmolyte concentrations of seagrass Posidonia australis. Mar. Pollut. Bull. 115, 252–260.

Del Pilar Ruso, Y., la Ossa Carretero, J.A.D., Casalduero, F.G., Lizaso, J.L.S., 2007. Spatial and temporal changes in infaunal communities inhabiting soft-bottoms affected by brine discharge. Mar. Environ. Res. 64, 492–503.

Del Pilar-Ruso, Y., De-la-Ossa-Carretero, J.A., Gimenez-Casalduero, F., Sanchez-Lizaso, J.L., 2008. Effects of a brine discharge over soft bottom polychaeta assemblage. Environ. Pollut. 156, 240–250.

de-la-Ossa-Carretero, J.A., Del-Pilar-Ruso, Y., Loya-Fernández, A., Ferrero-Vicente, L.M., Marco-Méndez, C., Martinez-Garcia, E., Sánchez-Lizaso, J.L., 2016. Response of amphipod assemblages to desalination brine discharge: impact and recovery. Estuar. Coast. Shelf Sci. 172, 13–23.

Del-Pilar-Ruso, Y., Martinez-Garcia, E., Giménez-Casalduero, F., Loya-Fernández, A., Ferrero-Vicente, L.M., Marco-Méndez, C., de-la-Ossa-Carretero, J.A., Sánchez-Lizaso, J.L., 2015. Benthic community recovery from brine impact after the implementation of mitigation measures. Water Res. 70, 325–336.

Drami, D., Yacobi, Y.Z., Stambler, N., Kress, N., 2011. Seawater quality and microbial communities at a desalination plant marine outfall. A field study at the Israeli Mediterranean coast. Water Res. 45, 5449–5462.

Dupavillon, J.L., Gillanders, B.M., 2009. Impacts of seawater desalination on the giant Australian cuttlefish Sepia apama in the upper Spencer Gulf, South Australia. Mar. Environ. Res. 67, 207–218.

Fernández-Torquemada, Y., Sánchez-Lizaso, J.L., 2005. Effects of salinity on leaf growth and survival of the Mediterranean seagrass Posidonia oceanica (L.) Delile. J. Exp. Mar. Biol. Ecol. 320, 57–63.

Fernández-Torquemada, Y., Sánchez-Lizaso, J., 2011. Responses of two Mediterranean seagrasses to experimental changes in salinity. Hydrobiologia 669, 21–33.

Fernández-Torquemada, Y., Sánchez-Lizaso, J.L., 2013. Effects of salinity on seed germination and early seedling growth of the Mediterranean seagrass Posidonia oceanica (L.) Delile. Estuar. Coast. Shelf Sci. 119, 64–70.

Fernandez-Torquemada, Y., Gonzalez-Correa, J.M., Loya, A., Ferrero, L.M., Diaz-Valdes, M., Sanchez-Lizaso, J.L., 2009. Dispersion of brine discharge from seawater reverse osmosis desalination plants. Desalin. Water Treat. 5, 137–145.

Frank, H., Rahav, E., Bar-Zeev, E., 2017. Short-term effects of SWRO desalination brine on benthic heterotrophic microbial communities. Desalination 417, 52–59.

Gacia, E., Invers, O., Manzanera, M., Ballesteros, E., Romero, J., 2007. Impact of the brine from a desalination plant on a shallow seagrass (Posidonia oceanica) meadow. Estuar. Coast. Shelf Sci. 72, 579–590.

Garrote-Moreno, A., Fernández-Torquemada, Y., Sánchez-Lizaso, J.L., 2014. Salinity fluctuation of the brine discharge affects growth and survival of the seagrass Cymodocea nodosa. Mar. Pollut. Bull. 81, 61–68.

Holloway, K., 2009. Perth Seawater Desalination Plant Water Quality Monitoring Programme. Final Programme summary Report 2005–2008. Report No. 445_001/3. Prepared by Oceanica Consulting Pty LTD for the Water Corporation of Western Australia.

Kämpf, J., Clarke, B., 2013. How robust is the environmental impact assessment process in South Australia? Behind the scenes of the Adelaide seawater desalination project. Mar. Policy 38, 500–506.

Koch, M.S., Schopmeyer, S.A., Kyhn-Hansen, C., Madden, C.J., Peters, J.S., 2007. Tropical seagrass species tolerance to hypersalinity stress. Aquat. Bot. 86, 14–24.

Kress, N., Galil, B.S., 2012. Seawater desalination in Israel and its environmental impact. Desal. Water Reuse 26–29 February-March 2012.

Kress, N., Shoham-Frider, E., Lubinevski, H., 2016. Marine monitoring at the brine outfalls of the Palmachim and Soreq desalination plants. Final report for the 2015 surveys. IOLR Report H12/2016 (In Hebrew).

Kress, N., Shoham-Frider, E., Lubinevski, H., 2017. Monitoring the Effect of Brine Discharge on the Marine Environment: A Case Study off Israel's Mediterranean Coast. The International Desalination Association World Congress on Desalination and Water Reuse 2017, Sao Paulo, Brazil.

Kupsco, A., Sikder, R., Schlenk, D., 2017. Comparative developmental toxicity of desalination brine and sulfate-dominated saltwater in a Euryhaline fish. Arch. Environ. Contam. Toxicol. 72, 294–302.

Le Page, S., 2005. Salinity tolerance investigations: a supplemental report for the Carlsbad, CA desalination project. Report presented to Poseidon Resources.
Lin, Y.-C., Chang-Chien, G.-P., Chiang, P.-C., Chen, W.-H., Lin, Y.-C., 2013. Potential impacts of discharges from seawater reverse osmosis on Taiwan marine environment. Desalination 322, 84–93.
Marín-Guirao, L., Sandoval-Gil, J.M., Ruíz, J.M., Sánchez-Lizaso, J.L., 2011. Photosynthesis, growth and survival of the Mediterranean seagrass Posidonia oceanica in response to simulated salinity increases in a laboratory mesocosm system. Estuar. Coast. Shelf Sci. 92, 286–296.
Marín-Guirao, L., Sandoval-Gil, J.M., Bernardeau-Esteller, J., Ruíz, J.M., Sánchez-Lizaso, J.L., 2013. Responses of the Mediterranean seagrass Posidonia oceanica to hypersaline stress duration and recovery. Mar. Environ. Res. 84, 60–75.
McConnell, R., 2009. Tampa Bay seawater desalination facility—environmental impact monitoring. In: Proceedings of 2009 Annual Water Reuse Conference, Seattle.
Mezhoud, N., Temimi, M., Zhao, J., Al Shehhi, M.R., Ghedira, H., 2016. Analysis of the spatio-temporal variability of seawater quality in the southeastern Arabian Gulf. Mar. Pollut. Bull. 106, 127–138.
Miri, M., Nabavi, S.M.B., Doustshenas, B., Safahieh, A.R., Loghmani, M., 2015. Levels of Heavy Metals in Sediments in the Vicinity of Chabahar Bay Desalination Plant.
Ozair, G., Al-Sebaie, K.Z., Al-Zahrany, S.A., 2017. Impact of Long Term Concentrated Brine Disposal on the Ecosystems of Nearshore Marine Environment—A Case Study. The International Desalination Association World Congress on Desalination and Water Reuse 2017, Sao Paulo, Brazil.
PBSJ, 2010. Tampa Bay Desalination Facility Hydrobiological Monitoring Data Summary 2002–2010. https://www.tampabaywater.org/documents/supplies/monitoring programs/surfacewatermonitoring/TampaBayDesalHBMPDataSummary2002-2010.pdf.
Portillo, E., Ruiz de la Rosa, M., Louzara, G., Ruiz, J.M., Marín-Guirao, L., Quesada, J., González, J.C., Roque, F., González, N., Mendoza, H., 2014. Assessment of the abiotic and biotic effects of sodium metabisulphite pulses discharged from desalination plant chemical treatments on seagrass (Cymodocea nodosa) habitats in the Canary Islands. Mar. Pollut. Bull. 80, 222–233.
Raventos, N., Macpherson, E., García-Rubiés, A., 2006. Effect of brine discharge from a desalination plant on macrobenthic communities in the NW Mediterranean. Mar. Environ. Res. 62, 1–14.
Riera, R., Tuya, F., Sacramento, A., Ramos, E., Rodriguez, M., Monterroso, O., 2011. The effects of brine disposal on a subtidal meiofauna community. Estuar. Coast. Shelf Sci. 93, 359–365.
Riera, R., Tuya, F., Ramos, E., Rodríguez, M., Monterroso, Ó., 2012. Variability of macrofaunal assemblages on the surroundings of a brine disposal. Desalination 291, 94–100.
Rivers, D., 2013. Perth desalination plant-Cockburn sound benthic macrofauna community and sediment habitat, repeat survey 2013. Oceanica Consulting. Report No. 604_01_006.
Ruiz, J.M., Marin-Guirao, L., Sandoval-Gil, J.M., 2009. Responses of the Mediterranean seagrass Posidonia oceanica to in situ simulated salinity increase. Bot. Mar. 52, 459–470.
Sadiq, M., 2002. Metal contamination in sediments from a desalination plant effluent outfall area. Sci. Total Environ. 287, 37–44.
Saeed, M.O., Al-Tisan, I.A., Ershath, M.I., 2017. Perspective on desalination discharges and coastal environments of the Arabian peninsula. The International Desalination Association World Congress on Desalination and Water Reuse 2017, Sao Paulo, Brazil: IDA17WC-58245.
Sánchez-Lizaso, J.L., Romero, J., Ruiz, J., Gacia, E., Buceta, J.L., Invers, O., Fernández Torquemada, Y., Mas, J., Ruiz-Mateo, A., Manzanera, M., 2008. Salinity tolerance

of the Mediterranean seagrass Posidonia oceanica: recommendations to minimize the impact of brine discharges from desalination plants. Desalination 221, 602–607.

Sandoval-Gil, J.M., Marin-Guirao, L., Ruiz, J.M., 2012a. Tolerance of Mediterranean seagrasses (Posidonia oceanica and Cymodocea nodosa) to hypersaline stress: Water relations and osmolyte concentrations. Mar. Biol. 159, 1129–1141.

Sandoval-Gil, J.M., Marín-Guirao, L., Ruiz, J.M., 2012b. The effect of salinity increase on the photosynthesis, growth and survival of the Mediterranean seagrass Cymodocea nodosa. Estuar. Coast. Shelf Sci. 115, 260–271.

Serra, I., Lauritano, C., Dattolo, E., Puoti, A., Nicastro, S., Innocenti, A., Procaccini, G., 2012. Reference genes assessment for the seagrass Posidonia oceanica in different salinity, pH and light conditions. Mar. Biol. 159, 1269–1282.

Shoham-Frider, E., Kress, N., Gordon, N., Lubinevski, H., 2017. Joint marine monitoring for Adama-Agan Ltd, Paz Refineries Ashdod Ltd, Ashdod Desalination Ltd. Final report for the 2016 surveys. IOLR Report H9/2017 (In Hebrew).

Shpir, D., Ben Yosef, D., 2017a. Monitoring the coastal and marine environment at the discharge site of the Orot Rabin power plant and the H2ID desalination plant. Results from 2016. Israel Electric Corp. RELP-3-2017.

Shpir, D., Ben Yosef, D., 2017b. Monitoring the coastal and marine environment at the discharge site of the Rutenberg power plant, VID desalination plant, Mekorot's well amelioration plant and Dorad's power plant. Results from 2016. Israel Electric Corp. RELP-21-2017.

Shute, S., 2009. Perth desalination plant-Cockburn sound benthic macrofauna community and sediment habitat, repeat macrobenthic survey. Oceanica consulting. Report No. 604-011/1: 202 pp.

Uddin, S., Al Ghadban, A.N., Khabbaz, A., 2011. Localized hyper saline waters in Arabian gulf from desalination activity-an example from South Kuwait. Environ. Monit. Assess. 181, 587–594.

van der Merwe, R., Hammes, F., Lattemann, S., Amy, G., 2014a. Flow cytometric assessment of microbial abundance in the near-field area of seawater reverse osmosis concentrate discharge. Desalination 343, 208–216.

van der Merwe, R., Röthig, T., Voolstra, C.R., Ochsenkühn, M.A., Lattemann, S., Amy, G.L., 2014b. High salinity tolerance of the Red Sea coral Fungia granulosa under desalination concentrate discharge conditions: an in situ photophysiology experiment. Front. Mar. Sci. 1.

Vars, S., Johnston, M., Hayles, J., Gascooke, J., Brown, M., Leterme, S., Ellis, A., 2013. 29Si {1H} CP-MAS NMR comparison and ATR-FTIR spectroscopic analysis of the diatoms Chaetoceros muelleri and Thalassiosira pseudonana grown at different salinities. Anal. Bioanal. Chem. 405, 3359–3365.

Vega, P.M., Artal, M.V., 2013. Impact of the discharge of brine on benthic communities: a case study in Chile. The International Desalination Association World Congress on Desalination and Water Reuse 2013, Tianjin, China: IDAWC/TIAN13-341.

Viskovich, P.G., Gordon, H.F., Walker, S.J., 2014. Busting a salty myth: Long-term monitoring detects limited impacts on benthic infauna after three years of brine discharge. IDA J. Desal. Water Reuse 6, 134–144.

Voutchkov, N., 2007. Assessing tolerance threshold of marine organisms to desalination plant discharges. Membrane Residuals Solutions.

FURTHER READING

Anon, 2016. Southern seawater desalination plant. Marine Environment Monitoring Annual Report 17 January 2015–16 January 2016Prepared by Water Corporation, Australia.

Carlsbad, 2017. https://ciwqs.waterboards.ca.gov/ciwqs/readOnly/PublicReportEsmrAtGlanceServlet?reportID=2&isDrilldown=true&documentID=1870868.

Hermony, A., Sutzkover-Gutman, I., Talmi, Y., Fine, O., 2014. Palmachim seawater desalination plant—seven years of expansions with uninterrupted operation together with process improvements. Desalin. Water Treat. 1–10.

Hobbs, D., Stauber, J., Kumar, A., Smith, R., 2008. Ecotoxicity of Effluent From the Proposed Olympic Dam Desalination Plant. Final Report. Hydrobiology Pty Ltd. Aquatic Environmental Services.

Kress, N., Galil, B., 2016. Impact of seawater desalination by reverse osmosis on the marine environment. In: Burn, S., Gray, S. (Eds.), Efficient Desalination by Reverse Osmosis. IWA, London, pp. 177–202.

Paul, P., Al Tenaiji, A., Braimah, N., 2016. A review of the water and energy sectors and the use of a nexus approach in Abu Dhabi. Int. J. Environ. Res. Public Health 13, 364.

Safrai, I., Zask, A., 2008. Reverse osmosis desalination plants—marine environmentalist regulator point of view. Desalination 220, 72–84.

UNEP. 2008. Desalination resource and guidance manual for environmental impact assessments. United Nations Environment Programme, Regional Office for West Asia, Manama, and World Health Organization, Regional Office for the Eastern Mediterranean, Cairo Ed. S. Lattemann: 168 pp.

CHAPTER 7

Policy and Regulations for Seawater Desalination

The environment and its natural resources are protected from anthropogenic impacts by legislation, regulation, and policies that set up requirements, criteria, and best environmental practice. They usually do not address desalination specifically, but are adapted regionally or locally to regulate desalination plants.

The scope of topics included in this chapter is wide, extensively described in the literature, and therefore presented here in a condensed way. They include legislation, regulations, different components that make up an environmental impact assessment process, mitigation measures, operational monitoring, and public engagement. Although explained in separate sections, most topics are intertwined and should be considered jointly, at times with the output of one element serving as the input to another.

This chapter is written with three objectives: (1) to expose the reader to the vast field of environmental law, environmental impact assessment, and environmental protection, emphasizing seawater desalination when possible; (2) to provide the means for further research and for adaptation to specific local conditions; and (3) to provide a brief outlook for each of the elements described.

The topics addressed here have been mentioned in the previous chapters without an in-depth explanation. Moreover, terms used and described in the previous chapters, such as marine food web, entrainment, impingement, and entrapment (EI&E), intake and discharge systems, are used across this chapter. They are not explained again and the reader is directed to the previous chapters and to the glossary.

7.1 BASIC CONCEPTS

Basic concepts, necessary for the understanding of this chapter, are explained in this section.

Marine Impacts of Seawater Desalination: Science, Management, and Policy
https://doi.org/10.1016/B978-0-12-811953-2.00007-4

The *near field* is the region where mixing is caused by turbulence and other processes generated by the discharge itself. These mechanisms entrain seawater that readily dilutes the effluent within a few minutes from the discharge and within tens to hundreds of meters from the discharge site. The effectiveness of the initial mixing is determined by the configuration of the discharge system, the volume and rate of discharge, and the density of effluent.

The *far field* is the region where mixing and dilution are caused by oceanic turbulence, currents, and stratification. The rate of mixing and dilution is much slower than in the near field (days to weeks) and occurs within hundreds of meters to kilometers from the outfall, highly dependent on the hydrographic conditions.

Emission limit values (*ELVs*), also known as effluent standards or discharge quality standards, are numerical values (or narrative statements) for effluent components at the point of discharge. Regulators and permitting authorities use ELVs as they are simple to administer, monitor, and enforce. ELVs encourage source control such as effluent treatment, recycling but do not consider directly the response of the receiving environment.

Environmental quality standards (*EQS*) or ambient standards are the ambient (or reference) concentrations permitted beyond the mixing zone. They directly consider the response of the receiving environment. EQS can be numerical values or narrative statements.

Mixing zone, a regulatory concept, is an allocated impact zone where EQS can be exceeded as long as acutely toxic conditions are prevented. EQS must be met at the edge of the mixing zone, and for toxic effluents, criteria must be met at the end of the pipe. For practical purposes, a mixing zone with a radius of 100 m is enough for a well-designed discharge system, but mixing zones from 50 to 800 m have been defined. A mixing zone is also referred to as the zone of initial dilution (ZID), and as an area of low ecological protection, among others.

Environmental impact assessment (*EIA*) is a systematic process by which the anticipated impacts of a proposed project are identified at the design and planning stages. Similar terms are found, at times with slightly different regulatory or procedural meaning, such as: environmental impact statement, environmental assessment, environmental review procedure, environmental impact study, strategic environmental assessment, ecological risk assessment, strategic environment, and social assessment. "Environmental impact" used in the scientific field pertains to the development of methodologies to identify and estimate effects on the environment, while in the legal administrative context it leads to the development of regulations for environmental protection.

Ecosystem approach (*EcAp*) is a strategy for the integrated management of land, water, and living resources that promotes conservation and sustainable use in an equitable way.

7.2 ENVIRONMENTAL LEGISLATION

Environmental law has become a major tool in environmental and natural resources management and in sustainable development. The Food and Agriculture Organization (FAO), the International Union for Conservation of Nature (IUCN), and the United Nations Environment Program (UNEP) developed and operate jointly the ECOLEX database. ECOLEX is a free, web-based, information service on treaties, international nonbinding soft-law, nonbinding policy, technical guidance documents, national legislation, judicial decisions, law, and policy literature concerning the environment.

Desalination is usually not addressed specifically in environmental legislation. Regulation of desalination plants is mostly derived from legislation pertaining to planning and building permanent structures, water intakes (such as power plant cooling waters), effluent's ELVs, EQS, and drinking water quality criteria. Those span from international environmental conventions and agreements, through regional treaties and conventions, to national and local legislation and regulation.

7.2.1 International Conventions and Agreements

The realization that environmental effects transcend national boundaries guides the institution of global environmental conventions, agreements, initiatives, and strategic objectives. Many are found under the auspices of the United Nations (UN) with relevant goals applicable to seawater desalination, even if not mentioned specifically. For example,

- The UN Conference on the Human Environment (also known as the Stockholm Conference, 1972) was the first major conference on international environmental issues and a turning point in the development of international environmental politics. The UN Conference for Environment and Development (the Earth summit) in Rio de Janeiro (1992) produced the Rio Declaration, known as Agenda 21, which helps countries create domestic and international environmental policies. Four of its principles are applicable to seawater desalination: Principle 17, Protection of the oceans, and all seas and adjacent coastal areas through integrated management and sustainable development; Principle 15, Conservation of biological diversity; Principle 8, Integrating environment and development in decision-making at the policy, planning, and management level; and Principle 12, Managing fragile ecosystems: combating desertification and drought. The principles and commitments of Agenda 21 were reaffirmed in Johannesburg 2002 (Rio + 10) and in Rio 2012 (Rio + 20).

- The UN Convention on Biological Diversity (CBD, 1992) established a target of 10% of the ocean to be protected by 2020. Seawater desalination may impact nearshore marine protected areas (MPAs) by reducing larval availability connectivity between MPAs due to EI&E and interfere with biodiversity conservation efforts.
- The UNEP hosts the secretariat for the Global Program of Action for the Protection of the Marine Environment from Land-based Activities (GPA) (Washington Declaration, 1995). Its common goal is the "sustained and effective action to deal with all land based impacts upon the marine environment, specifically those resulting from sewage, persistent organic pollutants, radioactive substances, heavy metals, oils (hydrocarbons), nutrients, sediment mobilization, litter, and physical alteration and destruction of habitat."
- The Convention on EIA in a Transboundary Context (Espoo, 1991) under the UN Economic Commission for Europe (UNECE) sets out the obligations of the parties to assess the environmental impact of activities at an early stage of planning. The parties should notify and consult with each other on projects that are likely to have a significant adverse environmental impact across boundaries. The strategic environmental assessment (SEA) protocol (Kyiv, 2003) augments the Espoo Convention
- In 2015, the UN member states adopted the 2030 Agenda for Sustainable Development and its 17 Sustainable Development Goals (SDGs). Among them, relevant for seawater desalination are: SDG 6, Clean water and sanitation to ensure availability and sustainable management of water and sanitation for all; and SDG 14, Life below water to conserve and sustainably use oceans, seas and marine resources. The FAO is assisting the countries to implement the Agenda.

Two additional international organizations addressing environmental protection, legislation, and management are the IUCN and the World Bank. The IUCN works with governments and local communities to develop and implement policies and actions based on the EcAp for *the conservation of nature and sustainable use*. The World Bank requires that the people and the environment are protected from potential adverse impact as a prerequisite to funding. The current environmental and social policies of the World Bank are known as the "Safeguard Policies," such as environmental and social impact assessments. Those will be incrementally replaced by a new set of environment and social policies called the Environmental and Social Framework.

7.2.2 Regional Agreements and Treaties

Regional agreements and treaties are usually based on global environmental conventions. Some are found in areas with extensive seawater desalination effort (The Gulf, Red Sea, and Mediterranean Sea) in the framework of several organizations. Some of the agreements are described in the following list.

- The Regional Seas Program implements many of UNEP's marine-related policies, addressing the degradation of the world's oceans and coastal areas. It engages neighboring countries in comprehensive and specific actions to protect their common marine environment in a "shared seas" approach.
  - Conventions administered by the UN. The Barcelona Convention for the protection of marine environment and the coastal region of the Mediterranean (adopted 1976) and implemented through the Mediterranean action plan (MAP). In 2003, UNEP-MAP published a report with an assessment and guidelines for seawater desalination in the Mediterranean Sea. The Cartagena Convention was adopted 1981 for the protection of the wider Caribbean region.
  - Two additional regional organizations not administered by the UN are the Regional Organization for the Protection of the Marine Environment (ROPME) of the Gulf States and Oman, established at the Kuwait Convention (adopted 1978) and the Regional Organization for the Conservation of the Environment of the Red Sea and Gulf of Aden (PERSGA), established at the Jeddah Convention (adopted 1982).
- The Environment Directorate General of the EU, set up in 1973, proposes policies and legislation that protect natural habitats using two legal acts: regulations and directives. Regulation is a binding legislative act that must be applied in its entirety across the EU and the directive, a legislative act that sets out goals that all EU countries must achieve. Although not explicitly addressing desalination, some of the directives may be applied to seawater desalination regulation:
  - The Water Framework Directive (WFD) 2000/60/EC sets critical limits on the discharged effluents or receiving water bodies in order to guarantee water protection.

  - The Marine Strategy Framework Directive (MSFD) 2008/56/EC aims to achieve good environmental status (GES) of the EU's marine waters by 2020 and to protect the resource base upon which marine-related economic and social activities depend.
  - Directive 2014/52/EU on the assessment of the effects of certain public and private projects on the environment (the EIA Directive). It defines the European normative framework on EIA and marks the realization of the prevention principle in environmental policy.
  - The Habitats Directive 92/43/EEC framework for the conservation of wild animal and plant species and natural habitats of community importance.
  - Directive 2008/105/EC and its amendment (2013/39/EC) on environmental quality standards in the field of water policy.
- The regional Australian and New Zealand Environment Conservation Council (ANZECC) published the ANZECC 2000 Guidelines for fresh and marine water quality that provide government, regulators, industry, consultants, community groups, and catchment and water managers with a framework for conserving ambient water quality in rivers, lakes, estuaries, and marine waters.

7.2.3 National Legislation and Regulations

On the national level, environmental legislation and regulation are often derived from global and regional agreements. They vary among different regions and are adapted to specific local environmental, economic, and social conditions leading to different prioritization of environmental problems. As with regional agreements, the national legislation often does not address desalination specifically but is adjusted to regulate it. In spite of the regional differences, local regulatory frameworks share some common requirements for seawater desalination: a permitting process and an EIA process at the planning stage, regulation for intake and discharge systems, ELVs and EQS criteria, product water quality criteria, marine monitoring during plant operations, and management measures to mitigate impacts. These topics are addressed at length in the subsequent sections. Frequently, the permitting process and regulatory enforcement fall within overlapping jurisdictions (ministries, commissions, permitting, and regulatory bodies) and complicates the legal process. Table 7.1 brings a few illustrative examples of national legislation and Table 7.2 shows guidelines for the salinity of brine discharged in the marine environment.

Table 7.1 Several illustrative examples of national environmental regulations applied to seawater desalination

Country/state	Requirements, planning stage	Requirements, operational stage	National regulation/guideline	Regional convention
United States, California	Permit, EIA, intake and discharge systems requirements, mitigation, WET	Monitoring, reporting, receiving water limitation for salinity Maximal intake velocity	Clean water act, National Pollutant Discharge Elimination system permit program, National Environmental Policy Act, California Ocean Plan	
Western Australia	Permit, EIA, environmental management plan	Monitoring, reporting, adherence to conditions of permit	Western Australia environmental protection act, Ministerial approval statement	ANZECC 2000 Guidelines
Southern Australia	Permit, EIA, modeling, toxicity testing	Minimum dilution required, mixing zone (100 m)	Section 46 of the Development Act 1993 Environmental Protection (Water Quality) Policy2003	
Sultanate of Oman	Permit for effluent discharge, EIA, license	Discharge limits and mixing zone (300 m)	Law for the Protection of Environment and Prevention of Pollution	ROPME
Kingdom of Saudi Arabia (special Commission for Jubail and Yanbu areas)	Permit, EIA (environmental and social)	Comply with guidelines in the discharge permit	The Public Environmental Law (Royal Decree No. M/34, 2001) Guidelines for concentrations in discharge and at the edge of the mixing zone (size on a case by case basis)	ROPME

Continued

Table 7.1 Several illustrative examples of national environmental regulations applied to seawater desalination—cont'd

Country/state	Requirements, planning stage	Requirements, operational stage	National regulation/guideline	Regional convention
Spain federal and autonomous regions legislation	EIA, permit, Record of Decision	Monitoring, reporting, action protocol if quality criteria exceeded, salinity guidelines for receiving waters	Royal Decree 1302/1986, Article 13 of the Consolidated Text of the Water Act (Royal Decree 1/2001) Law 21/2013	EU regulation (Habitats directive 92/43/EEC, Directive 2014/52/EU) Barcelona convention
Algeria			Law 05-12 (2005)	Barcelona convention
Egypt				PERSGA
Israel	EIA, permit, outfall system, pipeline construction	Monitoring, reporting, discharge quality standards	The Law for the Protection of the Coastal Environment, Law for the Prevention of Sea Pollution from Land-Based Sources	Barcelona convention
China		State and local discharge	Marine environmental protection law	
Chile	EIA	Effluent standards	Environmental norms, established in Supreme Decree N° 90/01	
Taiwan	EIA		Environmental impact assessment Act	UNEP

ANZECC, Australian and New Zealand Environment Conservation Council; *EIA*, environmental impact assessment; *PERSGA*, Regional Organization for the Conservation of the Environment of the Red Sea and Gulf of Aden; *ROPME*, Regional Organization for the Protection of the Marine Environment of the Gulf States and Oman; *UNEP*, United Nations environmental program; *WET*, whole effluent toxicity.

Table 7.2 A sample of regulations and recommendations for salinity and temperature limits for the discharge of desalination brine

Region/authority	Limit for salinity or % increase above ambient	Compliance point (relative to discharge)	Reference
US-EPA	≤1–4	50–300 m	Jenkins et al. (2012)[a]
California (CA), United States	<5%	Edge of regulatory mixing zone (100 m)	Jenkins and Wasyl (2014)
CA, United States	≤2.0	Edge of regulatory mixing zone (100 m)	California Ocean Plan (2015)
Carlsbad, CA, United States	≤40	305 m (1000 ft)	San Diego Regional Water Quality Control Board (2006)[a]
Huntington Beach, CA, United States	≤40	305 m (1000 ft)	Santa Ana Regional Water Quality Control Board (2012)[a]
Tampa, Florida, United States	Average salinity = 35.8		Water Reuse, white paper (2011)
Western Australia	<5%		Western Australia guidelines[a]
Oakajee Port, Western Australia	≤1		The Waters of Victoria State Environment Protection Policy[a]
Perth, Australia	≤1.2 ≤0.8	50 m 1000 m	Western Australia EPA[a]
Sydney, Australia	≤1	50–75 m	ANZECC (2000)
Gold Coast, Australia	≤2	120 m 60 m	GCD Alliance (2006)[a] Viskovich et al. (2014)
Southern SW desalination Australia	<1.3	LEPA boundary	Marine monitoring report
Adelaide, Australia	<1.3	1000 m	
Abu Dhabi (UAE)	≤5%	Mixing zone boundary	Jenkins et al. (2012)[a]
Sultanate of Oman	≤2	300 m	Sultanate of Oman (2005)[a]

Continued

Table 7.2 A sample of regulations and recommendations for salinity and temperature limits for the discharge of desalination brine—cont'd

Region/authority	Limit for salinity or % increase above ambient	Compliance point (relative to discharge)	Reference
Kuwait	<42		Uddin et al. (2011)
Spain	<38.5, 25% of the time <40, 5% of the time	Near *Posidonia oceanica* seagrass mats	Fuentes-Bargues (2014) and Palomar and Losada (2010)
Gaza	≤2 ≤0.5	100 m Edge of mixing zone	Abualtayef et al. (2016)
Okinawa, Japan	≤1	Mixing zone boundary	Okinawa Bureau for Enterprises

Region/authority	Temperature limit (°C), above ambient	Compliance point (relative to discharge)	Reference
US-EPA	<1	300 m	
World Bank	<3		
Australia, Barrup Peninsula	<2 <0.1	Outfall Edge of mixing zone	Ahmad and Baddour (2014)
Sultanate of Oman	<10 <1	End of pipe 300 m	Ahmad and Baddour (2014)
Israel, Ashkelon	<10		Discharge Permit for Ruthenberg power plant

[a]Reference cited from Jenkins et al. (2012).

7.3 ENVIRONMENTAL IMPACT ASSESSMENT

An EIA is a systematic process by which the anticipated impacts of a proposed project are identified at the design and planning stages (see Section 7.1), namely, a predictive process. EIA is a requirement in most international conventions and national regulations. The International Organization for Standardization (ISO) has published the standard ISO 14001:2015 for environmental management systems while UNEP published a comprehensive guidance manual in 2008 on EIA specific for desalination plants.

The EIA provides information on the environmental and social consequences of a project to promote environmentally sound and sustainable

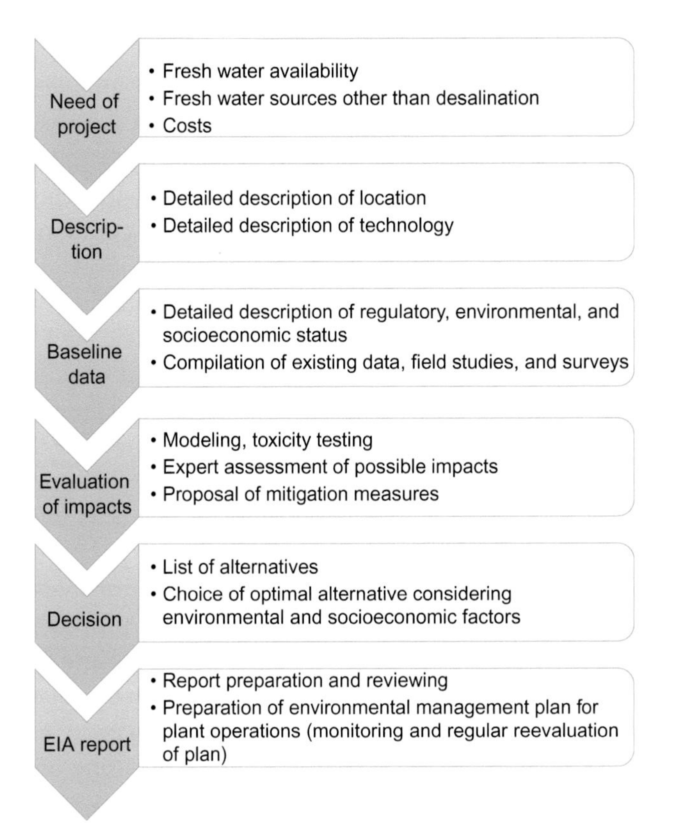

Fig. 7.1 A succinct depiction of an environmental impact assessment process for seawater desalination. Public engagement is recommended during all stages of the process.

development through the identification and choice of appropriate alternatives and mitigation measures. EIAs are planned according to the requirements of the local regulatory body and have ultimately to be approved by it. Each EIA process is unique, contingent on the project specifics and on the location of the desalination plant. A succinct general depiction of the process is found in Fig. 7.1. Below are the general guidelines for an EIA process, at times performed simultaneously. The EIA process and report should incorporate:

- The purpose and need of the desalination project, including availability and costs of alternative water sources (water treatment and reuse, water conservation, prevention of water waste).

- Social sustainability. Impacts on human health (desalinated water quality), land use, population growth, infrastructure, confidence in desalinated water supply, impact on recreational activities, or other legitimate uses of the sea and the coastline.
- Project description: proposed location, co-location with other industries or marine uses, the onshore and offshore physical components of the plant (buildings, pumps, pipelines, intake, and brine discharge systems), planned construction activities and timeline, connection to the drinking water supply grid.
- Technology description: desalination process, engineering specifications, production capacity, energy source and usage, intake and discharge systems, pretreatment of source water (coagulation, biocide application, antiscaling measures, cleaning stages, desalinated water treatment), type, volume and composition of discharges and emissions (marine, terrestrial and atmospheric).
- Environmental baseline description: Compilation and analysis of existing data on the terrestrial and marine habitats at the proposed location, performance of baseline monitoring surveys prior to construction.
- Modeling: Loss of organism's EI&E (see Chapter 4) at the intake systems, local (near and far field) hydrography and brine dispersion, transboundary transport, and impact on seawater quality and marine organisms.
- Toxicity testing of discharges.
- Assessment of possible impacts.
- Decision among alternatives: Environmental risk assessment and multicriteria decision analysis tools to define and choose the optimal alternative and set up mitigation measures.
- Description of measures to be undertaken in order to avoid or mitigate negative impacts during the construction phase and at the desalination plant operational phase, considering the following:
  - Best available technology (BAT). The latest stage of development (state of the art) of processes, facilities, or methods of operation that indicate the practical suitability of a particular measure for limiting discharges, emissions, and waste.
  - Best environmental practice (BEP). The application of the most appropriate combination of environmental control measures and strategies.
  - The precautionary principle. Action should be taken to prevent serious adverse impacts, even if no proof exists, only suggestive evidence of effect.

Lately, an additional step to the EIA is envisaged to account for the effect of changing climate. Within desalination, it could be the increase for freshwater demand; increase in temperature and salinity of the feed water (seawater); and increase in phytoplankton blooms.

7.4 MODELING

A model is a conceptual or mathematical simplification used, among others, to investigate a real natural system, a risk assessment problem, and/or a decision-making process. Modeling is a common requirement in an EIA process and a basic element for well-informed decision-making and regulatory processes. It is a diverse field continuously expanding, with dedicated professional societies, conferences, and numerous publications in scientific journals and books. Models can be commercial, developed by private companies, or open source (available free) developed usually by environmental agencies and the academia.

Models used in engineering and plant design and to describe the natural systems before and during plant operations (environmental perspective) are classified as hydrodynamic (or physical), biogeochemical (or water quality, ecological), and ecosystem based models (see Fig. 7.2) They describe mathematically the state of a dynamic system, using state variables and the processes connecting among them. As one moves from the physics to the ecosystem, the natural processes are more complex, include more variables and processes. Many are poorly known, therefore assumptions, best guesses, and simplifications are inherent in modeling. A compromise must be reached between maximizing details and parameterization with high uncertainty and simple models that may lack important processes necessary for science and management. Therefore, a model should be carefully selected, based on its capabilities and limitations, and the EIA requirements. From the management perspective, environmental risk assessment and multi-criteria decision analysis (MCDA) models are used to choose the optimal conditions (environmental, economic, social) for the desalination plant.

All models need to be initiated, calibrated, verified, and validated. This is done by estimating the parameters and processes of the natural and socioeconomic systems, setting up its starting conditions, checking consistency of the results and if they reproduce real-life observations. A sensitivity analysis is also performed to check how and to what extent the results change due to variations in the model's parameters. Error and uncertainty are estimated and should be a part of the EIA report.

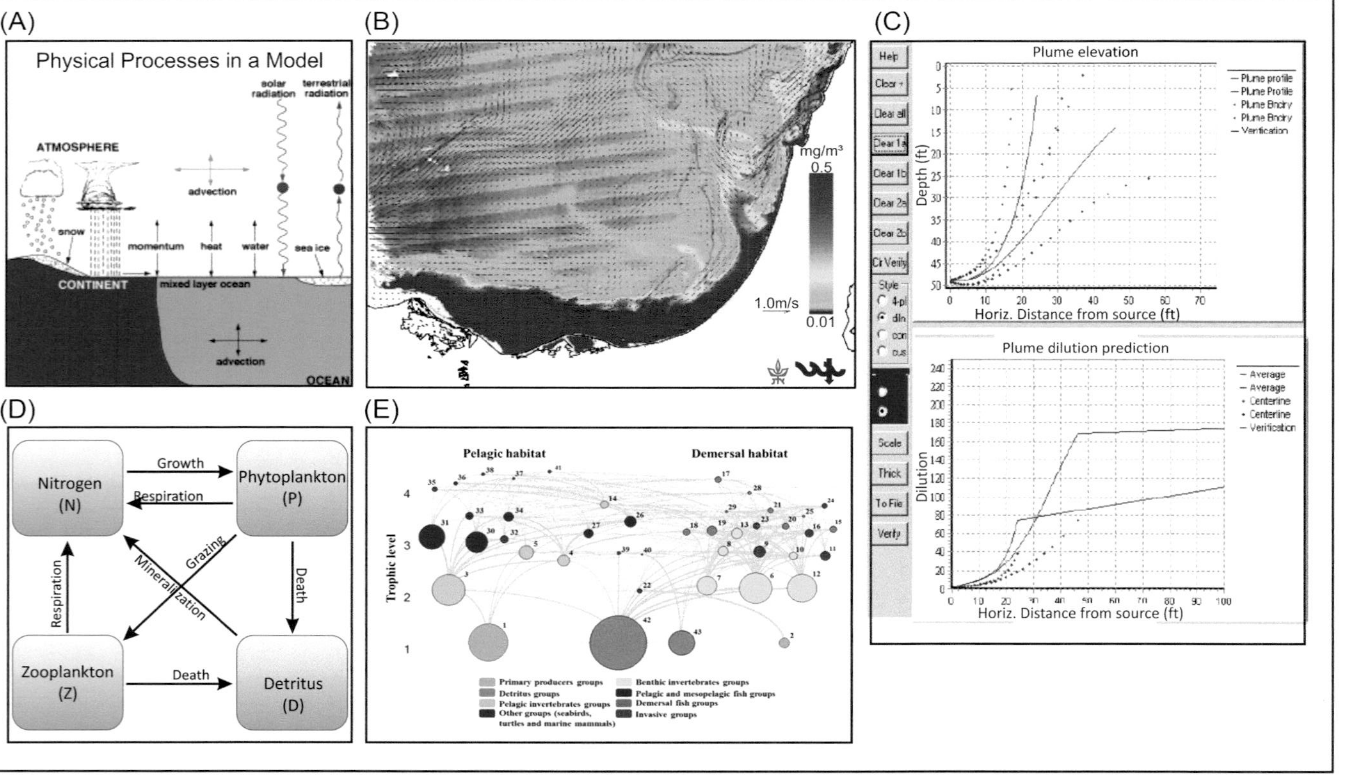

Fig. 7.2 Representation of different models used in an EIA. (A) Parameters included in hydrodynamic modeling. (B) Output of an hydrodynamic model showing wave height superimposed with a satellite image of chlorophyll *a*. (C) Output of a near field hydrodynamic model. (D) Schematics of a simple NPZD biogeochemical model. (E) Ecopath's ecological model flow diagram. *From (A) NOAA; (B) https://isramar.ocean.org.il/isramar2009/cosem/fsle.aspx; (C) Visual Plumes Manual; (E) Corrales, X., Ofir, E., Coll, M., Goren, M., Edelist, D., Heymans, J.J., Gal, G., 2017. Modeling the role and impact of alien species and fisheries on the Israeli marine continental shelf ecosystem. J. Mar. Syst. 170, 88–102.*

Still, with all these caveats in place, modeling is a very useful tool and a crucial aspect of planning and operation (brine management) of a desalination plant. The equations and in-depth methodology of modeling are not discussed in this section; however, it provides a general picture, describes the tools available for modeling, and directs the reader to the vast literature on this topic.

7.4.1 Hydrodynamic Modeling

Hydrodynamic models are the most widespread, common, and robust among the models describing natural systems. Hydrodynamic models simulate conservative parameters, such as temperature and salinity, which change as a result of physical processes only and not due to chemical and biological processes. Model outputs include, among others, strength and direction of currents, temperature and salinity fields, sea water levels, sea surface elevations, and tides.

Existing models (both commercial and open source) must be adapted to local conditions using specific meteorological data (temperature, heat exchange), oceanographic data (currents, T, S, sea surface elevation), bottom bathymetry, and desalination plant characteristics. Use of these models for the planning of desalination plants usually emphasizes worst-case scenarios to simulate maximal impact. One such scenario is maximal brine discharge into calm seas, with slow currents and mixing, and therefore low and slow dispersion of the effluents.

The wide range of time and distance of near- and far-field processes (see Section 7.1), makes it necessary to use separate models to describe their specific processes. However, following the separate simulations, near- and far-field models results are coupled to obtain an overall description of brine plume behavior.

7.4.1.1 Near-Field Modeling

Near-field models are used to predict the dynamics and initial dilution of an effluent (i.e., brine) as it exits the discharge system (see Section 7.1). Some commonly used near-field models are:

- CORMIX (Cornell Mixing Zone Expert). The CORMIX is a commercial model for the near field and approved by the US Environmental Protection Agency (EPA). The model is used as a hydrodynamic mixing zone model and as a decision support system for analyzing and predicting pollutants' initial dispersion after discharge into the marine environments. It includes several modules such as the CorJet and CorSens.

- VISUAL PLUMES. The visual plumes model is a free, Windows-based, application developed by the EPA. It simulates the mixing of effluents following marine discharge and is used also for ocean outfall design, analysis of mixing zone, and water quality. It includes several modules such as the UM3 module.
- VISJET. Innovative Modeling and Visualization Technology for Environmental Impact Assessment) software is a commercial model developed by the University of Hong Kong, which can simulate positively and negatively buoyant discharges. It includes several modules such as the JetLag module.

In addition to the mathematical models, *physical or experimental modeling* can be used to simulate the near field. Physical modeling consists of laboratory experiments using scaled down discharge systems that simulate the specific case being tested. Tests can be carried out on any effluent, discharge configuration, and ambient conditions. Physical modeling is particularly useful where mathematical models are not verified or uncertain.

7.4.1.2 Far-Field Modeling

Far-field models are used to predict the fate and transport of coastal discharges as it is transported by currents and density-driven flow following the initial dilution in the near field. They may be two dimensional (2D, depth-averaged), adequate for fairly shallow and nonstratified waters. However, in most cases, three-dimensional (3D) models, that describe the depth distribution of the variables, are needed. These models usually incorporate a module for sediment transport, important in planning and during the construction of desalination plants.

Far-field hydrodynamic models require extensive field data for set up, calibration, and validation. Frequently used far-field models include:

- Delft3D from Deltares. The Delft3D is a commercial far-field hydrodynamic 3D model to investigate hydrodynamics, sediment transport and morphology, and water quality for coastal environments. Three of its modules are available as open source.
- MIKE 3 from DHI. The MIKE 3 model is a commercial hydrodynamic 3D model used to investigate, among others, mixing and dispersion of marine discharges and coastal circulation. MIKE 21 is a similar model but only 2D (depth-averaged). Both versions of the model include several modules.
- General Ocean Circulation Models. These 3D hydrodynamic models were usually developed by and for the scientific community. They

include, among others, the Princeton Ocean Model (POM), Regional Ocean Modeling System (ROMS), Coupled Hydrodynamical Ecological model for Regional Shelf Seas (COHERENS).

7.4.2 Biogeochemical Modeling

Biogeochemical modeling is used to model seawater quality, the fate of substances, the link between chemical components and organisms, and biological processes at the lower trophic levels (bacteria, phytoplankton). In contrast to the state variables in hydrodynamic modeling, biogeochemical variables are nonconservative, changing as a result of chemical and biological processes such as oxidation, photosynthesis, and remineralization in addition to changes due to physical processes.

Biogeochemical modeling lags behind hydrodynamics in terms of usage and robustness and has a higher uncertainty. Initially, biogeochemical models were developed to answer relatively simple questions related to: carbon fluxes, the response of water column variables (nutrient, oxygen, and chlorophyll-a concentrations) to changes in nutrient loads and hydrology. A well-known 1D model is the NPZD (nitrate, phytoplankton, zooplankton, detritus) model (see Fig. 7.2). Considerably more complex models are now in use and in development due to the improvement in computational resources, increased understanding and parametrization of natural processes, increased resolution and quality of observational data, and improved coupling to the hydrodynamic models.

Biogeochemical models simulate, among others, dissolved oxygen concentrations, nutrient cycling, phytoplankton growth, eutrophication, exchange of substances with the atmosphere and the sediments, adsorption and desorption to particles. Some commonly used biogeochemical models are:

- D-Water Quality (D-WAQ) from Deltares, a multidimensional water quality model framework. D-WAQ is not a hydrodynamic model, so the physical information is derived from Delft3D output.
- MIKE ECO Lab from DHI is numerical modeling tool for ecological modeling integrated in both MIKE 21 (2D) and MIKE 3 (3D).
- The Water Quality Analysis Simulation Program (WASP) by the EPA is a dynamic compartment-modeling program, including both the water column and the benthos. WASP allows the user to investigate 1D, 2D, and 3D systems, and a variety of pollutant types.

Specific biogeochemical models, of variable complexity, are also developed and used from scientific research of local and global processes.

7.4.3 Ecosystem-Based Modeling

Ecosystem-based models aim at understanding the structure and functioning of ecosystems and at quantifying their trophic flows, emphasizing higher-trophic-level organisms. They are usually called end-to-end models, combining physicochemical processes with low and high trophic level descriptors and interactions. Therefore, ecosystem based models are data extensive with high uncertainty. There is growing interest in these models due to increased demand for quantitative tools to support the EcAp toward environmental management to achieve and maintain GES. They are used to project and analyze the consequences of anthropogenic changes on the food web structure and function, in particular for fisheries management. Ecosystem models could be used to evaluate the impact of seawater intake and brine discharge from seawater desalination on the different trophic levels of the marine ecosystem.

A well-known example of an ecosystem model is the Ecopath with Ecosim (EwE). Originally developed by scientists from the National Oceanic and Atmospheric Administration (NOAA), it is now centered at Ecopath International Initiative. The site has a compilation of the projects that used EwE. EwE is a free ecological/ecosystem modeling software with three main components: Ecopath, Ecosim, and Ecospace. Ecopath is a static, mass-balanced snapshot of the system and their interactions, represented by trophically linked biomass "pools" of single species or species groups. It needs as input biomass, production, consumption, ecotrophic efficiency for each group. Ecosim is a time dynamic simulation module with initial parameters calculated by Ecopath. Ecospace is a spatial and temporal dynamic module primarily designed for exploring impact and placement of protected areas. An EwE model requires the collection, compilation, and harmonization of information derived from observations and computer simulations: species abundance, biomass trends, diet, life history, species production, consumption, and ecosystem properties.

NOAA has also a fisheries Virtual Ecosystem Scenario Viewer (VES-V) that visually illustrates the responses of marine ecosystems to a range of living marine resource management scenarios.

Two additional ecosystem-based models are: Atlantis and OSMOSE, but more can be found in the literature (i.e., MS-PROD, NEMURO, SEAPOYM):

- Atlantis, developed by scientists of the Commonwealth Scientific and Industrial Research Organization (CSIRO, Australia). The model tracks the flow of nutrients through the trophic levels, using nitrogen as the "common currency" but also carbon and phosphorous. Oceanographic processes, such as water fluxes, salinity, and temperature conditions are obtained from hydrographic models or from observational data. The main ecological processes considered are production, consumption and predation, waste production and cycling, migration, reproduction and recruitment, habitat dependency, and mortality.

- OSMOSE, developed by the Exploited Marine Ecosystem unit funded by the French national institute for research and sustainable development (IRD). This is a multispecies and individual-based model that focuses on fish species. It assumes opportunistic predation based on spatial co-occurrence and size adequacy between a predator and its prey (size-based opportunistic predation).

7.4.4 Entrainment Modeling

Modeling biotic loss due to entrainment at a seawater intake system is a special case of ecosystem based model. It is not commonly used in seawater desalination operations but has been used extensively for seawater intake of cooling waters to coastal power plants in the United States, California in particular. Concisely, these models require a vast knowledge on: plant operations, local and regional hydrography (obtained from hydrodynamic modeling), numbers and identity of the plankton entrained, and their life history and natural mortality to assess the following information:

- Adult equivalent loss—equates the number of fish eggs and larvae lost as a result of intake operation to an equivalent number of adults of the same species.
- Fecundity hindcast—calculates the area of fecund females required to offset the number of eggs and larvae lost as a result of intake operation
- Area production foregone—calculates the area of spawning habitat required to offset the number of eggs and larvae lost as a result of intake operation. In California, entrainment impacts are commonly compensated for by creating or restoring a fish habitat at a nearby location to produce the organisms lost at the intake.

Impingement impacts were also modeled in similar ways for power plants. However, impingement in a seawater desalination plant can be mitigated easily by introducing physical and engineering constraints such as reducing intake velocity and installing velocity caps at the intake (see Chapter 4).

7.4.5 Risk Assessment and Decision-Making Models

Risk assessment and decision making are the last steps of an EIA. They are tightly interconnected and receive inputs from other models and elements of the EIA process.

7.4.5.1 Environmental Risk Assessment

In general, risk assessment is a systematic process to formulate a problem and evaluate and organize data, information, assumptions, and uncertainties. It aims to understand and predict the relationships between stressors and effects and to estimate the probability of harm using statistics and mathematical tools.

For environmental risk assessment, a well-known approach follows the driver-pressure-state-impact-response (DPSIR) framework. DPSIR assumes a chain of causal links starting with driving forces (anthropogenic activities and processes), through pressures (direct stresses from the anthropogenic activities), to states (physical, chemical, and biological environmental conditions) and impacts (measure of effects on ecosystems and humans due to changes in state) leading to responses (management actions to reduce impacts). Establishing a DPSIR framework is a complex task as all the various cause-effect relationships have to be carefully described and environmental changes can rarely be attributed to a single cause. Fig. 7.3 depicts a possible DPSIR diagram for seawater desalination.

7.4.5.2 Decision Making

An MCDA sets and recommends decision rules on how to evaluate the overall environmental, technical, and socioeconomic performance of a set of alternatives and how to choose among them. The steps of an MCDA process are to: define the problem and formulate objectives, find decision alternatives and predict the impacts of each, rank the alternatives and decide. The implementation of the decision is accompanied by regulatory, legal and economic processes, by stakeholders involvement, and by evaluation of the correctness of the decision by following the real responses after the implementation of the project.

MCDA is a vast field of research, with its scientific community and its specialized journals, as well as a large and growing number of real-world applications. The analytic hierarchy process (AHP) is an MCDA method used to derive ratio scales from paired comparisons. The input can be obtained from actual measurements or from subjective opinions such as

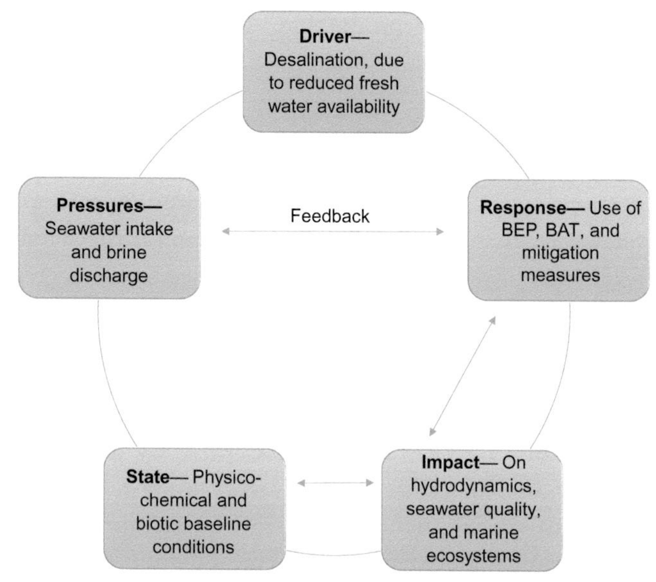

Fig. 7.3 A schematic representation of a possible driver-pressure-state-impact-response (DPSIR) framework for seawater desalination.

satisfaction and preference. For example, AHP has been used in the context of seawater desalination in Kuwait (for sustainable water security strategy); in Taiwan (to explore the EIA commissioner's perceptions of the impact of a desalination project); in the United States to examine stakeholders' priorities for the development of seawater desalination facilities.

Additional MCDA methods are data envelopment analysis, analytical network process, multiattribute utility theory, among others. Software packages are available for use (i.e., PROMETHEE, ELECTRE, TOPSIS). The actual MCDA method should be chosen and adapted based on the type of the decision problem formulation.

7.5 TOXICITY TESTING

Toxicity testing compares the response of an organism exposed to a specific chemical to the response of the unexposed, same species, organism under identical conditions. Acute toxicity is determined by exposure to high concentrations for a limited time, and chronic toxicity is determined by exposure to lower concentrations for longer times.

Toxicity tests are performed with different organisms at different life stages, and the responses examined are mortality and sublethal effects (i.e., changes in behavior, reproduction, growth, enzymatic processes, appearance, willingness to feed, activity, weight gain/loss). These tests are used to derive specific concentrations, such as: lowest observed effect concentrations (LOEC), no observed effect concentrations (NOEC), EC_{50}—concentration that causes an effect on 50% of the population, IC_{50}—concentration that causes an inhibition of growth of 50% in unicellular algae bioassay. A special case of toxicity testing is the whole effluent toxicity (WET) test in which organisms are exposed to the aggregate toxic effect from all components present in an effluent.

A combination of modeling results with acute and chronic toxicity data are used to derive "safe dilutions" in a species-sensitive distribution. It is followed by a statement such as "to protect 9X% of the species the effluent should be diluted by a factor of Y" where X and Y are site-specific values.

Regulatory bodies, such as the EPA and the European Commission, and international organizations, such as the ASTM International and the Organization for Economic Cooperation and Development (OECD) try to standardize the toxicity tests to be intercomparable by developing, publishing, or adopting consensus technical standards. Examples of toxicity tests and how to perform them are detailed in these sources.

Some examples of toxicity testing performed within the context of seawater desalination are presented in Chapter 6. An adaptation of WET has also been used to check the biotic effects of desalination brine. Briefly, locally important test organisms, representative of different groups (i.e., fish, invertebrate, macro-algae) are exposed to desalination brine at concentrations derived from modeling. The acute toxicity test establishes if the organism will survive the extreme conditions that may occur within the ZID, under worst-case scenario, and retain its capacity to reproduce. The chronic test tracks how well the organism can handle a long-term, steady state, exposure to the average concentration simulated for the ZID.

In the future, toxicity tests will strive to combine established procedures using whole organisms with in vitro assays and computational approaches to predict toxicity on the basis of an understanding of the biological processes at the molecular level.

7.6 PUBLIC ENGAGEMENT

A large-scale project, such as a desalination plant, may affect the daily life in a community, and impact natural resources and local and regional development.

Therefore, an EIA should engage diverse stakeholders to evaluate the project from its social and public aspects during most stages of the process (see Fig. 7.1). Public engagement is important for assembling different ideas and opinions, to understand concerns, and to conceive solutions to prevent or resolve conflicts prior to the execution of a project. The participatory process is performed by means of questionnaire-based surveys, public meetings, workshops, and comments on EIA reports.

Specifically, seawater desalination development entails potential public impacts such as increase in water price, reduced public access to coastal areas, and disruption of recreational activities. Studies of public position toward desalination fall into three contexts: general view toward desalination as a water supply option, opinions about actual projects, and attitudes in locations where desalination has already been implemented.

Listening and responding to public concern and a transparent decision-making process are very important. Skepticism about the need of a desalination plant or its environmental impact may delay and even cancel its implementation. Three short examples follow. The Western Galilee desalination plant was planned to be the fourth large-scale desalination plant along the Mediterranean Coast of Israel. Five large-scale plants are in full operation along the coast in 2018 while the Western Galilee plant is still at the planning stages due to extensive public opposition to the siting of the plant. In Australia, the desalination plant proposed for Point Lowly attracted a lot of public concern as it was to be located adjacent to an aggregation and spawning area for the iconic giant Australian cuttlefish, *Sepia apama*. The building of the plant was deferred. In California, there is a strong public opposition to seawater desalination based on water cost, energy usage, greenhouse gases emissions, and habitat impact. They advocate for water conservation, wastewater purification, stormwater capture, and brackish water desalination prior to the implementation of seawater desalination.

7.7 ENVIRONMENTAL MONITORING

A general definition of monitoring is the systematic, repeated measurement of biotic and abiotic parameters of the environment with a predefined spatial and temporal design. It applies equally to aquatic, terrestrial, and atmospheric environment. Monitoring, as modeling, is a usual EIA requirement during a plant's preconstruction and construction stages and a regulatory requirement during plant operations. Monitoring is site-specific, based on the local environmental setting and the plant's technology and configuration. It is usually prepared in accordance with national regulation and guidelines and requires the approval of relevant regulatory agencies prior to implementation.

The purpose of preconstruction monitoring is to describe the natural abiotic and biotic state using relevant parameters and determine their natural spatial and temporal variations. It serves as a baseline (reference) to estimate the magnitude of the impact at the operational phase. Baseline monitoring usually follows a stressor-based approach in which the potential stressors, the affected environmental receptors, and pathways for interaction are identified.

During the construction phase, monitoring is set up to follow the possible impacts identified at the EIA. Most impacts are localized and cease after the construction phase. However, some may be significant during construction, and the monitoring should assess if the inflicted impact is acceptable and recommend any necessary mitigation measures.

The main monitoring effort takes place at the operational phase of the desalination plant, called the operational monitoring. It aims to quantify and assess the accuracy of the possible impacts predicted during the EIA and to detect new impacts, if any. It also aims at ensuring that the regulatory requirements and quality standards are being met (compliance monitoring). Most of the actual data on the effects of seawater desalination on the marine environment were collected within the framework of compulsory operational monitoring (see Chapter 6).

Operational monitoring usually follows both the stressor-based approach and the effects-based approach. The latter measures the environmental state of the ecosystem by comparing environmental indicators from the impact site to the same indicators from a reference site, to account for the natural temporal and spatial variability. This is the so-called BACI approach, that is, "Before and After" and "Control and Impact", extensively used and documented in the literature.

Specific considerations and requirements for an effective operational monitoring are detailed in the following list:

- Operational monitoring should follow the baseline survey performed during the EIA, but not be restricted by it, and be in line with the permitting conditions for plant operations. In the case of seawater desalination, all marine compartments should be assessed (seawater, sediment, and biota both in seawater and in the sediment).

- It should be a long-term commitment, at least throughout the lifetime of the desalination plant. Long-term data series, with proper controls, are essential to account for natural temporal variability and to prevent erroneous conclusions on the environmental effects.
- Sampling frequency and methods should be based on site-specific characteristics. For seawater desalination the sampling stations should cover the impact area (within the mixing zone), affected area (beyond the mixing zone but still under the influence of the brine), and reference areas (where no brine is present).
- Sampling and measurement methodology should be accurate and precise to allow for the identification of small changes over the natural variability.
- The incorporation of remote sensing (sea surface temperature, color, and more) and geographic information system (GIS) to analyze, manage, and present spatial data is recommended.
- Operational monitoring should be adaptive. New or previously unknown pressures may emerge and existing pressures may decrease or increase. Changes in the desalination process may change brine discharge volume, salinity, and chemical composition.
- New tools and methodologies should be introduced to improve the monitoring program, for example, underwater autonomous vehicles with sensors, in situ chemical and optical sensors, and genomic and transcriptomics tools for biological research.
- Well-documented quality assurance and quality control protocols should be in place to ensure reliable and reproducible results across regions and time.
- Monitoring data should be analyzed regularly and critically to: allow for changes in the monitoring design when needed, enforce permitting license requirements, and require mitigation steps when effects are deemed excessive.
- A detailed monitoring report, including survey description, sampling and analysis methods, results, discussion, and recommendations should be issued. It is desirable to publish and disseminate the results to elicit feedback among regulators, plant operators, and scientists.

In addition to operational monitoring, in-plant monitoring is often required during plant operations. It includes monitoring the quality of the product water, the water quality of the source water (seawater intake), the composition of the brine discharged into the marine environment, and often toxicity testing of the brine.

REFERENCES

Abualtayef, M., Al-Najjar, H., Mogheir, Y., Seif, A.K., 2016. Numerical modeling of brine disposal from Gaza central seawater desalination plant. Arab. J. Geosci. 9, 572.
Ahmad, N., Baddour, R.E., 2014. A review of sources, effects, disposal methods, and regulations of brine into marine environments. Ocean Coast. Manag. 87, 1–7.
Fuentes-Bargues, J.L., 2014. Analysis of the process of environmental impact assessment for seawater desalination plants in Spain. Desalination 347, 166–174.
Jenkins, S.A., Wasyl, J., 2014. Brine dilution analysis for deep water desal, LLC, Monterey Bay regional water project at Moss Landing, CA. Report to Deep Water Desal, LLC.
Jenkins, S., Paduan, J., Roberts, P., Schlenk, D., Weis, J., 2012. Management of Brine Discharges to Coastal Waters. Recommendations of a Science Advisory Panel. Technical Report 694, Southern California Coastal Water Research Project, Costa Mesa, CA.
Palomar, P., Losada, I.J., 2010. Desalination in Spain: Recent developments and recommendations. Desalination 255, 97–106.
Uddin, S., Al Ghadban, A.N., Khabbaz, A., 2011. Localized hyper saline waters in Arabian Gulf from desalination activity-an example from South Kuwait. Environ. Monit. Assess. 181, 587–594.
Viskovich, P.G., Gordon, H.F., Walker, S.J., 2014. Busting a salty myth: long-term monitoring detects limited impacts on benthic infauna after three years of brine discharge. IDA J. Desal. Water Reuse 6, 134–144.

FURTHER READING

Abdul-Wahab, S.A., Jupp, B.P., 2009. Levels of heavy metals in subtidal sediments in the vicinity of thermal power/desalination plants: a case study. Desalination 244, 261–282.
Alharbi, O.A., Phillips, M.R., Williams, A.T., Gheith, A.M., Bantan, R.A., Rasul, N.M., 2012. Desalination impacts on the coastal environment: Ash Shuqayq, Saudi Arabia. Sci. Total Environ. 421–422, 163–172.
Aliewi, A., El-Sayed, E., Akbar, A., Hadi, K., Al-Rashed, M., 2017. Evaluation of desalination and other strategic management options using multi-criteria decision analysis in Kuwait. Desalination 413, 40–51.
Allen, J.I., Somerfield, P.J., Gilbert, F.J., 2007. Quantifying uncertainty in high-resolution coupled hydrodynamic-ecosystem models. J. Mar. Syst. 64, 3–14.
Allison, A.E.F., Dickson, M.E., Fisher, K.T., Thrush, S.F., 2018. Dilemmas of modelling and decision-making in environmental research. Environ. Model. Softw. 99, 147–155.
Al-Sharrah, G., Lababidi, H.M.S., Al-Anzi, B., 2017. Environmental ranking of desalination plants: the case of the Arabian gulf. Toxicol. Environ. Chem. 99, 1054.
Ayala, V., Kildea, T., Artal, J., 2015. Adelaide desalination plant-environmental impact studies. The International Desalination Association World Congress on Desalination and Water Reuse 2015, San Diego, CA, USA, IDAWC15-Ayala_51444.
Baawain, M., Choudri, B., Ahmed, M., Purnama, A., 2015. Recent Progress in Desalination, Environmental and Marine Outfall Systems. Springer International Publishing.
Barau, A.S., Al Hosani, N., 2015. Prospects of environmental governance in addressing sustainability challenges of seawater desalination industry in the Arabian Gulf. Environ. Sci. Pol. 50, 145–154.
Barnthouse, L.W., 2013. Impacts of entrainment and impingement on fish populations: a review of the scientific evidence. Environ. Sci. Pol. 31, 149–156.
Belatoui, A., Bouabessalam, H., Hacene, O.R., De-La-Ossa-Carretero, J.A., Martinez-Garcia, E., Sanchez-Lizaso, J.L., 2017. Environmental effects of brine discharge from

two desalination plants in Algeria (South Western Mediterranean). Desalin. Water Treat. 76, 311–318.
Bruggeman, J., Bolding, K., 2014. A general framework for aquatic biogeochemical models. Environ. Model. Softw. 61, 249–265.
Cooley, H., Ajami, N., Heberger, M., 2013. Key Issues in Seawater Desalination in California. Marine Impacts. Pacific Institute Report.
Corrales, X., Ofir, E., Coll, M., Goren, M., Edelist, D., Heymans, J.J., Gal, G., 2017. Modeling the role and impact of alien species and fisheries on the Israeli marine continental shelf ecosystem. J. Mar. Syst. 170, 88–102.
Drouiche, N., Ghaffour, N., Naceur, M.W., Mahmoudi, H., Ouslimane, T., 2011. Reasons for the fast growing seawater desalination capacity in Algeria. Water Resour. Manag. 25, 2743–2754.
El-Sadek, A., 2010. Water desalination: an imperative measure for water security in Egypt. Desalination 250, 876–884.
Falkenberg, L.J., Styan, C.A., 2015. The use of simulated whole effluents in toxicity assessments: a review of case studies from reverse osmosis desalination plants. Desalination 368, 3–9.
Ganju, N.K., Brush, M.J., Rashleigh, B., Aretxabaleta, A.L., del Barrio, P., Grear, J.S., Harris, L.A., Lake, S.J., McCardell, G., O'Donnell, J., Ralston, D.K., Signell, R.P., Testa, J.M., Vaudrey, J.M.P., 2016. Progress and challenges in coupled hydrodynamic-ecological estuarine modeling. Estuar. Coasts 39, 311–332.
Giwa, A., Dufour, V., Al Marzooqi, F., Al Kaabi, M., Hasan, S.W., 2017. Brine management methods: Recent innovations and current status. Desalination 407, 1–23.
Gude, V.G. (Ed.), 2018. Sustainable Desalination Handbook. Plant Selection, Design and Implementation. Elsevier Inc.
Heck, N., Paytan, A., Potts, D.C., Haddad, B., Petersen, K.L., 2017. Management priorities for seawater desalination plants in a marine protected area: a multi-criteria analysis. Mar. Policy 86, 64–71.
Hocking, G., 2012. Seawater desalination: an environmental regulator's perspective. Desalin. Water Treat. 51, 273–279.
Ishizaka, A., Nemery, P., 2013. Multi-Criteria Decision Analysis: Methods and Software. John Wiley & Sons, Ltd.
Kämpf, J., Clarke, B., 2013. How robust is the environmental impact assessment process in South Australia? Behind the scenes of the Adelaide seawater desalination project. Mar. Policy 38, 500–506.
Khan, S.J., Murchland, D., Rhodes, M., Waite, T.D., 2009. Management of concentrated waste streams from high-pressure membrane water treatment systems. Crit. Rev. Environ. Sci. Technol. 39, 367–415.
Khordagui, H., 2013. Assessment of potential cumulative environmental impacts of desalination plants around the Mediterranean Sea. SWIM Final Report, Activity 1.3.2.1.
Kupsco, A., Sikder, R., Schlenk, D., 2017. Comparative developmental toxicity of desalination brine and sulfate-dominated saltwater in a Euryhaline fish. Arch. Environ. Contam. Toxicol. 72, 294–302.
Latorre, M., 2005. Environmental impact of brine disposal on Posidonia seagrasses. Desalination 182, 517–524.
Lattemann, S., Amy, G., 2012. Marine monitoring surveys for desalination plants—A critical review. Desalin. Water Treat. 51, 233–245.
Lior, N., 2017. Sustainability as the quantitative norm for water desalination impacts. Desalination 401, 99–111.
Liu, T.-K., Sheu, H.-Y., Tseng, C.-N., 2013. Environmental impact assessment of seawater desalination plant under the framework of integrated coastal management. Desalination 326, 10–18.

Liu, T.-K., Weng, T.-H., Sheu, H.-Y., 2018. Exploring the environmental impact assessment commissioners' perspectives on the development of the seawater desalination project. Desalination 428, 108–115.
Mackey, E.D., Pozos, N., Wendle, J., Seacord, T., Hunt, H., Mayer, D.L., 2011. Assessing seawater intake systems for desalination plants. Water Research Foundation, Denver, CO.
Mansour, S., Arafat, H.A., Hasan, S.W., 2017. Brine Management in Desalination Plants. In: Desalination Sustainability. Elsevier, pp. 207–236 (Chapter 5).
Mickley, M.M., 2006. Membrane Concentrate Disposal: Practices and Regulation. Desalination and water purification research and development program No. 123 (second edition). U.S. Department of Interior.
Missimer, T.M., Maliva, R.G., 2018. Environmental issues in seawater reverse osmosis desalination: intakes and outfalls. Desalination 434, 198–215.
Missimer, T.M., Jones, B., Maliva, R.G. (Eds.), 2015. Intakes and Outfalls for Seawater Reverse-Osmosis Desalination Facilities: Innovations and Environmental Impacts. Springer International Publishing, Cham.
National Academies of Sciences, Engineering, and Medicine, 2017. Using 21st Century Science to Improve Risk-Related Evaluations. The National Academies Press, Washington, DC. https://doi.org/10.17226/24635.
Navarro, T., 2018. Water reuse and desalination in Spain—challenges and opportunities. J. Water Reuse Desal. 8, 153–168.
NRC, 2007. Toxicity Testing in the 21st Century: A Vision and a Strategy. National Research Council of the National Academies. The National Academies press, Washington, DC.
Nunes-Vaz, R.A., 2012. The salinity response of an inverse estuary to climate change & desalination. Estuar. Coast. Shelf Sci. 98, 49–59.
Palomar, P., Losada, I.J., 2011. Impacts of brine discharge on the marine environment. Modelling as a predictive tool. In: Schorr, M. (Ed.), Desalination, Trends and Technologies.
Palomar, P., Lara, J.L., Losada, I.J., Rodrigo, M., Alvárez, A., 2012. Near field brine discharge modelling part 1: analysis of commercial tools. Desalination 290, 14–27.
Purnama, A., Shao, D., 2015. Modeling brine discharge dispersion from two adjacent desalination outfalls in coastal waters. Desalination 362, 68–73.
Rose, K.A., Allen, J.I., Artioli, Y., Barange, M., Blackford, J., Carlotti, F., Cropp, R., Daewel, U., Edwards, K., Flynn, K., Hill, S.L., HilleRisLambers, R., Huse, G., Mackinson, S., Megrey, B., Moll, A., Rivkin, R., Salihoglu, B., Schrum, C., Shannon, L., Shin, Y.-J., Smith, S.L., Smith, C., Solidoro, C., St. John, M., Zhou, M., 2010. End-to-end models for the analysis of marine ecosystems: challenges, issues, and next steps. Mar. Coast. Fish. 2, 115–130.
Sadhwani, J.J., Veza, J.M., Santana, C., 2005. Case studies on environmental impact of seawater desalination. Desalination 185, 1–8.
Safrai, I., Zask, A., 2008. Reverse osmosis desalination plants—marine environmentalist regulator point of view. Desalination 220, 72–84.
Sale, P.F., Feary, D.A., Burt, J.A., Bauman, A.G., Cavalcante, G.H., Drouillard, K.G., Kjerfve, B., Marquis, E., Trick, C.G., Usseglio, P., Van Lavieren, H., 2011. The growing need for sustainable ecological Management of Marine Communities of the Persian Gulf. Ambio 40, 4–17.
Seip, K.L., Wenstop, F., 2006. A Primer on Environmental Decision-Making. An Integrative Quantitative Approach. Springer, The Netherlands.
Spiritos, E., Lipchin, C., 2013. Desalination in Israel. In: Becker, N. (Ed.), Water Policy in Israel: Context, Issues and Options. Springer Netherlands, Dordrecht, pp. 101–123.
Szeptycki, L., Hartge, E., Ajami, N., Erickson, A., Heady, W.N., LaFeir, L., Meister, B., Verdone, L., Koseff, J.R., 2016. Marine and Coastal Impacts on Ocean Desalination in California. Dialogue Report Compiled by Water in the West, Center for Ocean Solutions, Monterey Bay Aquarium and The Nature Conservancy, Monterey, CA.

UNEP, 2008. Desalination resource and guidance manual for environmental impact assessments. United Nations Environment Programme, Regional Office for West Asia, Manama, and World Health Organization, Regional Office for the Eastern Mediterranean, Cairo Ed. S. Lattemann, 168 pp.
UNEP/MAP/MEDPOL, 2003. Sea Water Desalination in the Mediterranean: Assessment and Guidelines. MAP Technical Reports Series no. 139 UNEP/MAP, Athens.
van der Merwe, R., Lattemann, S., Amy, G., 2012. A review of environmental governance and its effects on concentrate discharge from desalination plants in the Kingdom of Saudi Arabia. Desalin. Water Treat. 51, 262–272.
Vega, P.M., Artal, M.V., 2013. Impact of the discharge of brine on benthic communities: a case study in Chile. The International Desalination Association World Congress on Desalination and Water Reuse 2013, Tianjin, China, IDAWC/TIAN13-341.
Voutchkov, N., 2010. Seawater desalination: US desalination industry addresses obstacles to growth. Filtr. Sep. 47, 36–39.
Voutchkov, N., 2011. Overview of seawater concentrate disposal alternatives. Desalination 273, 205–219.
WHO, 2007. Desalination for safe water supply. Guidance for the health and environmental aspects applicable to desalination. World Health Organization, WHO/SDE/WSH/07.

INDEX

Note: Page numbers followed by *f* indicate figures and *t* indicate tables.

L

M

Printed in the United States
By Bookmasters